INORGANIC CHEMISTRY

FOURTH EDITION

GARY L. MIESSLER • DONALD A. TARR
St. Olaf College Northfield, Minnesota

Prentice Hall
Boston Columbus Indianapolis New York San Francisco Upper Saddle River
Amsterdam Cape Town Dubai London Madrid Milan Munich Paris Montréal Toronto
Delhi Mexico City São Paulo Sydney Hong Kong Seoul Singapore Taipei Tokyo

Editor in Chief, Chemistry: Adam Jaworski
Marketing Manager: Erin Gardner
Assistant Editor: Carol DuPont
Marketing Assistant: Nicola Houston
Managing Editor, Chemistry and Geosciences: Gina M. Cheselka
Project Manager: Wendy A. Perez
Supplement Cover Designer: Seventeenth Street Studios
Operations Specialist: Maura Zaldivar

Pearson Prentice Hall
Pearson Education, Inc.
Upper Saddle River, New Jersey 07458

Printed in the United States of America
10 9 8 7 6 5 4 3 2 1

ISBN-13: 978-0-13-612867-0
ISBN-10: 0-13-612867-X

Prentice Hall
is an imprint of

www.pearsonhighered.com

Contents

Preface

2 Atomic Structure 1

3 Simple Bonding Theory 15

4 Symmetry and Group Theory 28

5 Molecular Orbitals 42

6 Acid-Base and Donor-Acceptor Chemistry 59

7 The Crystalline Solid State 68

8 Chemistry of the Main Group Elements 76

9 Coordination Chemistry I: Structures and Isomers 88

10 Coordination Chemistry II: Bonding 99

11 Coordination Chemistry III: Electronic Spectra 112

12 Coordination Chemistry IV: Reactions and Mechanisms 122

13 Organometallic Chemistry 128

14 Organometallic Reactions and Catalysis 145

15 Parallels between Main Group and Organometallic Chemistry 157

16 Bioinorganic and Environmental Chemistry 166

Contents

Preface

Atomic Structure

Simple Bonding Theory

Symmetry and Group Theory

Molecular Orbitals

Acid-Base and Donor-Acceptor Chemistry

The Crystalline Solid State

Chemistry of the Main Group Elements

Coordination Chemistry I: Structures and Isomers

Coordination Chemistry II: Bonding

Coordination Chemistry III: Electronic Spectra

Coordination Chemistry IV: Reactions and Mechanisms

Organometallic Chemistry

Organometallic Reactions and Catalysis

Parallels Between Main Group and Organometallic Chemistry

Bioinorganic and Environmental Chemistry

Preface

This manual was prepared with the objective that it be useful for both students and instructors. Our general practice at St. Olaf College has been to provide students with answers to problems after they have attempted to work out their own solutions. However, we recognize that for some students, particularly those who are studying inorganic chemistry independently, the availability of a complete answer key can be a great help, and we have written this manual with such use also in mind.

Many problems in the text provide suggested literature references. In addition to making use of information from these sources, some solutions provide additional references, especially from the recent literature. Readers are encouraged to consult both types of references when possible, especially if they need additional details beyond those provided here.

The fourth edition of *Inorganic Chemistry* has more than 115 new problems, many with multiple parts. In addition, for this Solutions Manual the entire text of the preceding edition has been typed anew, and all diagrams are new or have been redrawn. Consequently, it is likely that careful readers will find places to offer suggestions for improvements. Such suggestions, either for the *Inorganic Chemistry* text or for this manual, are very welcome and can help improve future printings and editions.

Special thanks is due to St. Olaf student Kaitlin Hellie for generating new orbital images, retyping the manuscript from the third edition, and formatting the diagrams within the text. And appreciation is especially due to Carol DuPont for all she has done in her role as editor to shepherd this project through to its conclusion.

We hope that you will find this a useful resource.

Gary L. Miessler

CHAPTER 2: ATOMIC STRUCTURE

2.1 **a.** $\lambda = \dfrac{h}{mv} = \dfrac{6.626\times10^{-34}\text{ J s}}{9.110\times10^{-31}\text{kg}\times0.1\times2.998\times10^{8}\text{ms}^{-1}} = 2.426\times10^{-11}\text{ m}$

b. $\lambda = \dfrac{h}{mv} = \dfrac{6.626\times10^{-34}\text{ J s}}{0.400\text{kg}\times(10\text{km/hr}\times10^{3}\text{m/km}\times1\text{hr/3600s})} = 6\times10^{-34}\text{m}$

c. $\lambda = \dfrac{h}{mv} = \dfrac{6.626\times10^{-34}\text{ J s}}{8.0\text{ lb}\times0.4536\text{ kg/lb}\times2.0\text{ m s}^{-1}} = 9.1\times10^{-35}\text{ m}$

d. $\lambda = \dfrac{h}{mv} = \dfrac{6.626\times10^{-34}\text{ J s}}{13.7\text{ g}\times\text{kg}/10^{3}\text{g}\times30.0\text{ mi hr}^{-1}\times1\text{ hr/3600s}\times1609.3\text{ m/mi}}$

$= 3.61\times10^{-33}\text{ m}$

2.2 $E = R_H\left(\dfrac{1}{2^2}-\dfrac{1}{n_h^2}\right)$; $R_H = 1.097\times10^{7}\text{m}^{-1} = 1.097\times10^{5}\text{cm}^{-1}$; $\lambda = \dfrac{hc}{E} = \dfrac{1}{\bar{\nu}}$

$n_h = 4$ $\quad E = R_H\left(\dfrac{1}{4}-\dfrac{1}{16}\right) = R_H\left(\dfrac{12}{64}\right) = 20{,}570\text{cm}^{-1} = 4.085\times10^{-19}\text{ J}$

$\lambda = 4.862\times10^{-5}\text{cm} = 486.2\text{ nm}$

$n_h = 5$ $\quad E = R_H\left(\dfrac{1}{4}-\dfrac{1}{25}\right) = R_H\left(\dfrac{21}{100}\right) = 23{,}040\text{cm}^{-1} = 4.577\times10^{-19}\text{ J}$;

$\lambda = 4.341\times10^{-5}\text{cm} = 434.1\text{ nm}$

$n_h = 6$ $\quad E = R_H\left(\dfrac{1}{4}-\dfrac{1}{36}\right) = R_H\left(\dfrac{8}{36}\right) = 24{,}380\text{cm}^{-1} = 4.841\times10^{-19}\text{ J}$;

$\lambda = 4.102\times10^{-5}\text{cm} = 410.2\text{ nm}$

2.3 $E = R_H\left(\dfrac{1}{4}-\dfrac{1}{49}\right) = R_H\left(\dfrac{45}{196}\right) = 25{,}190\text{cm}^{-1} = 5.002\times10^{-19}\text{ J}$

$\lambda = \dfrac{1}{\bar{\nu}} = 3.970\times10^{-5}\text{cm} = 397.0\text{ nm}$

2.4 386.95 nm $\quad E = \dfrac{hc}{\lambda} = \dfrac{(6.626\times10^{-34}\text{ Js})(2.998\times10^{8}\text{m/s})}{(383.65\text{nm})(\text{m}/10^{9}\text{nm})} = 5.178\times10^{-19}\text{ J}$

379.90 nm $\quad E = \dfrac{hc}{\lambda} = \dfrac{(6.626\times10^{-34}\text{ Js})(2.998\times10^{8}\text{m/s})}{(379.90\text{nm})(\text{m}/10^{9}\text{nm})} = 5.229\times10^{-19}\text{ J}$

$$E = R_H\left(\frac{1}{2^2}-\frac{1}{n_h^{\;2}}\right); \quad \frac{1}{n_h^{\;2}} = \frac{1}{4}-\frac{E}{R_H} \text{ and } n_h = \left(\frac{1}{4}-\frac{E}{R_H}\right)^{-\frac{1}{2}}$$

For 386.95 nm: $n_h = \left(\frac{1}{4}-\frac{5.178\times10^{-19}J}{2.1787\times10^{-18}J}\right)^{-\frac{1}{2}} = 9$

For 379.90 nm: $n_h = \left(\frac{1}{4}-\frac{5.229\times10^{-19}J}{2.1787\times10^{-18}J}\right)^{-\frac{1}{2}} = 10$

2.5 The least energy would be for electrons falling from the $n = 4$ to the $n = 3$ level:

$$E = R_H\left(\frac{1}{3^2}-\frac{1}{4^2}\right) = 2.1787\times10^{-18}J\left(\frac{7}{144}\right) = 1.059\times10^{-19}J$$

The energy of the electromagnetic radiation emitted in this transition is too low for humans to see, in the infrared region of the spectrum.

2.6 **a.** $-\frac{h^2}{8\pi^2 m}\frac{\partial^2\Psi}{\partial x^2} = E\,\Psi; \quad \Psi = A\sin rx + B\cos sx$

$$\frac{\partial\Psi}{\partial x} = Ar\cos rx - Bs\sin sx$$

$$\frac{\partial^2\Psi}{\partial x^2} = -Ar^2\sin rx - Bs^2\cos sx \quad = \frac{h^2}{8\pi^2 m}\left(Ar^2\sin rx + Bs^2\cos sx\right)$$

$$= E\left(A\sin rx + B\cos sx\right)$$

If this is true, then the coefficients of the sine and cosine terms must be independently equal:

$$\frac{h^2Ar^2}{8\pi^2 m} = EA; \quad \frac{h^2Bs^2}{8\pi^2 m} = EB$$

$$r^2 = s^2 = \frac{8\pi^2 mE}{h^2}; \quad r = s = \sqrt{2mE}\,\frac{2\pi}{\text{h}}$$

b. $\Psi = A\sin rx$; when $x = 0$, $\Psi = A\sin 0 = 0$
when $x = a$, $\Psi = A\sin ra = 0$

$$\therefore ra = \pm n\pi; \; r = \pm\frac{n\pi}{a}$$

c. $E = \frac{r^2h^2}{8\pi^2 m} = \frac{n^2\pi^2}{a^2}\frac{h^2}{8\pi^2 m} = \frac{n^2h^2}{8ma^2}$

d. $$\int_0^a \Psi^2 dx = \int_0^a A^2 \sin^2\left(\frac{n\pi x}{a}\right) dx = A^2 \frac{a}{n\pi} \int_0^a \sin^2\left(\frac{n\pi x}{a}\right) d\left(\frac{n\pi x}{a}\right)$$

$$= \frac{a}{n\pi} A^2 \left[\frac{1}{2}\left(\frac{n\pi x}{a}\right) - \frac{1}{4}\sin\left(\frac{2n\pi x}{a}\right)\right]_0^a = 1$$

$$= \frac{aA^2}{n\pi}\left[\frac{1}{2}\frac{n\pi a}{a} - \frac{1}{4}\sin 2n\pi - 0 + \frac{1}{4}\sin 0\right] = 1$$

$$A = \sqrt{\frac{2}{a}}$$

2.7 **a.** $3p_z$ $4d_{xz}$

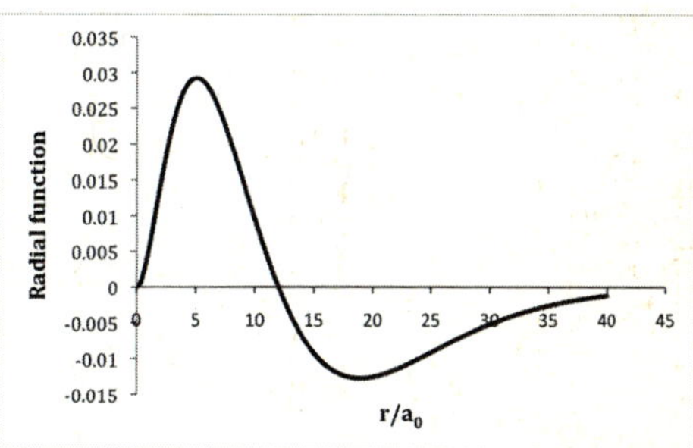

b. $3p_z$ $4d_{xz}$

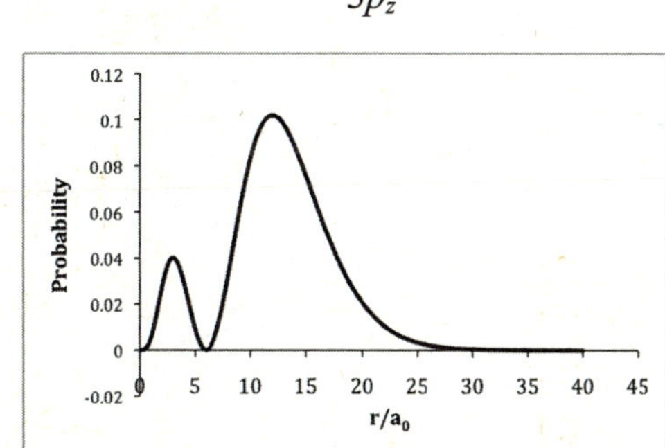

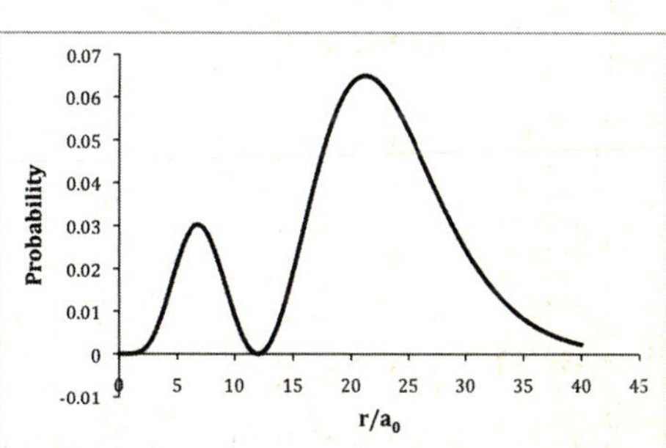

c. $3p_z$ $4d_{xz}$

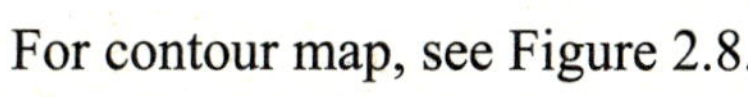

For contour map, see Figure 2.8.

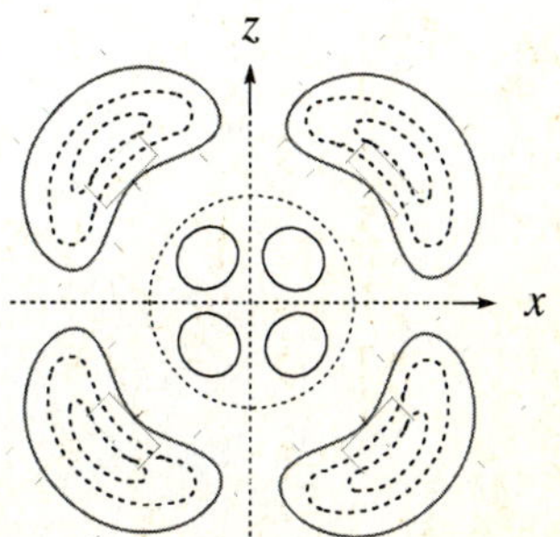

2.8 **a.** $4s$ $5d_{x^2-y^2}$

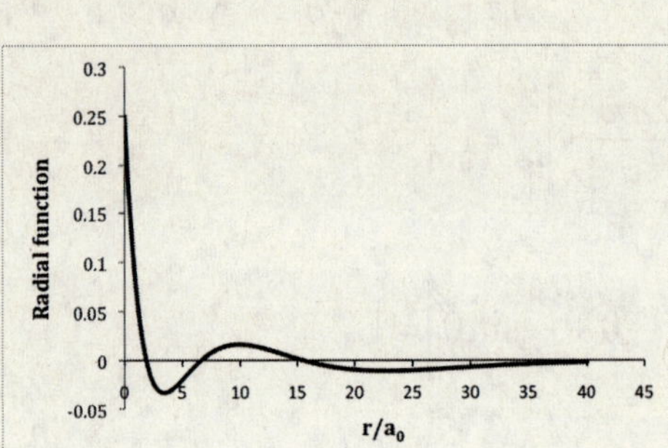

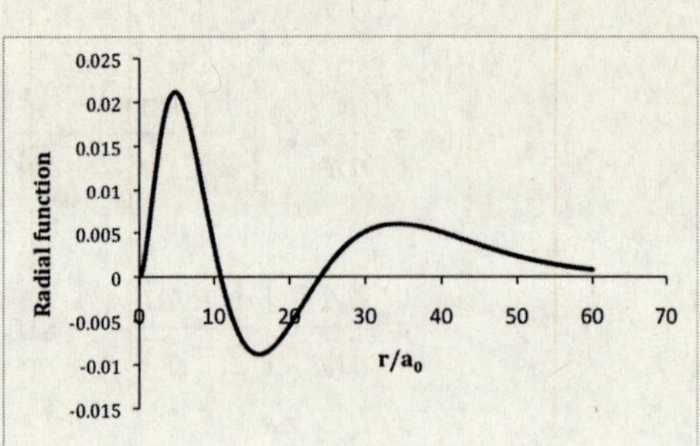

b. $4s$ $5d_{x^2-y^2}$

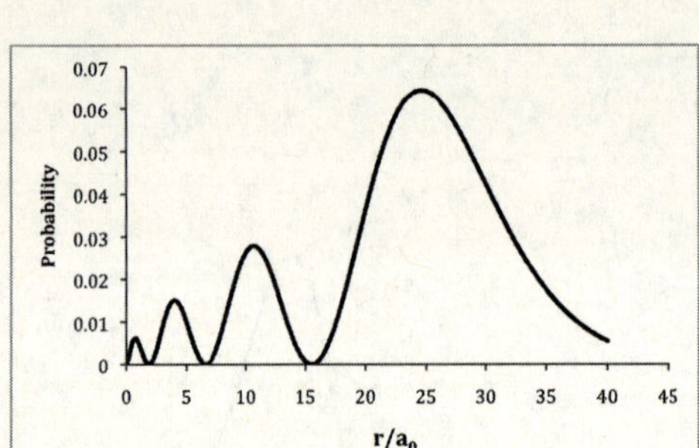

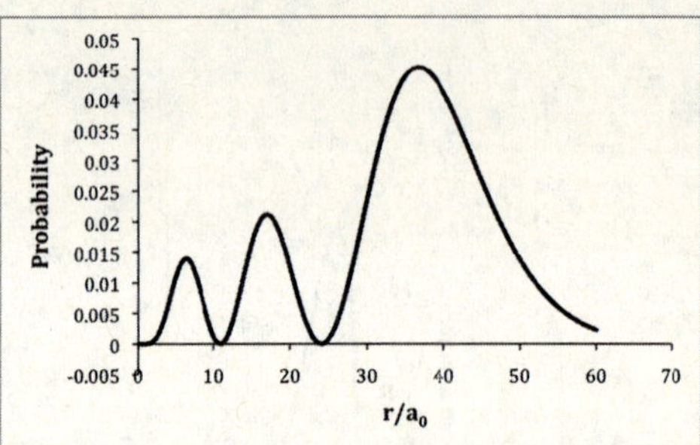

c. $4s$ $5d_{x^2-y^2}$

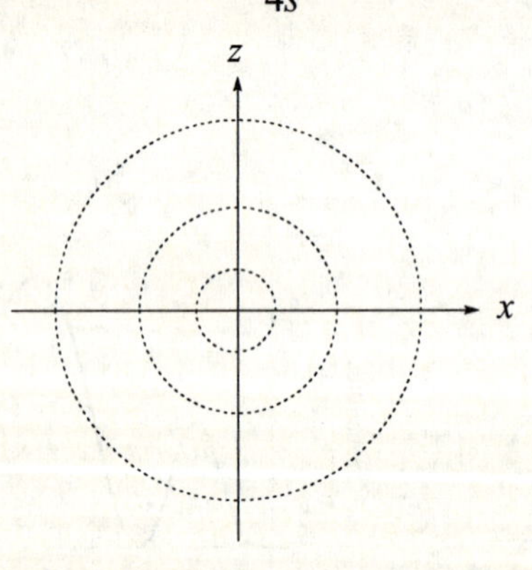

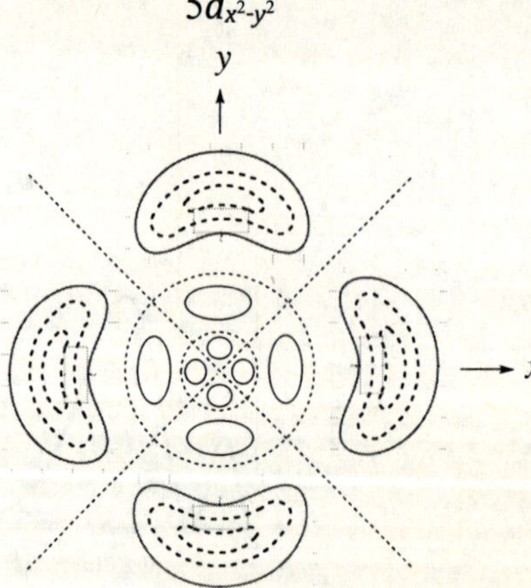

Dashed lines represent radial nodes. Electron density is inside the smallest node, between the nodes, and outside the largest node.

2.9 **a.**

5s | $4d_{z^2}$

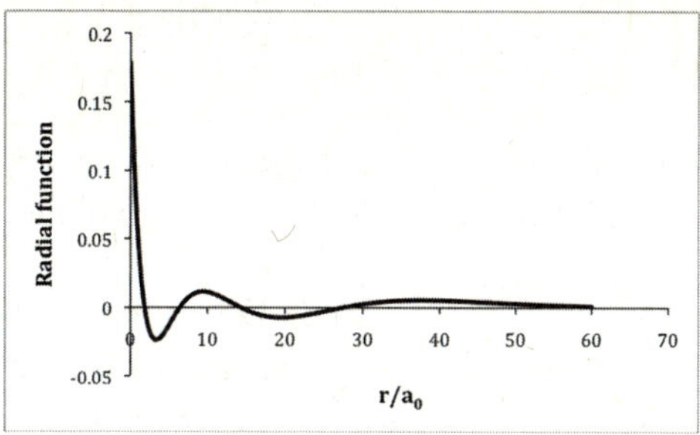

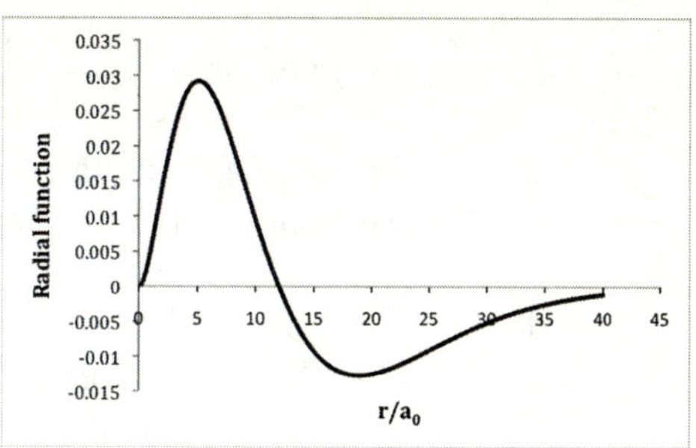

b.

5s | $4d_{z^2}$

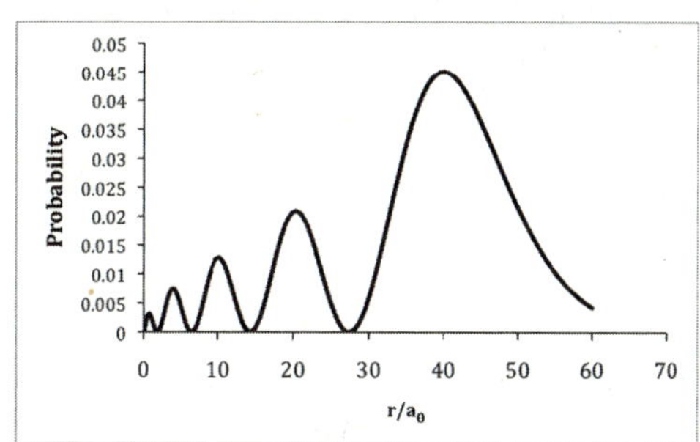

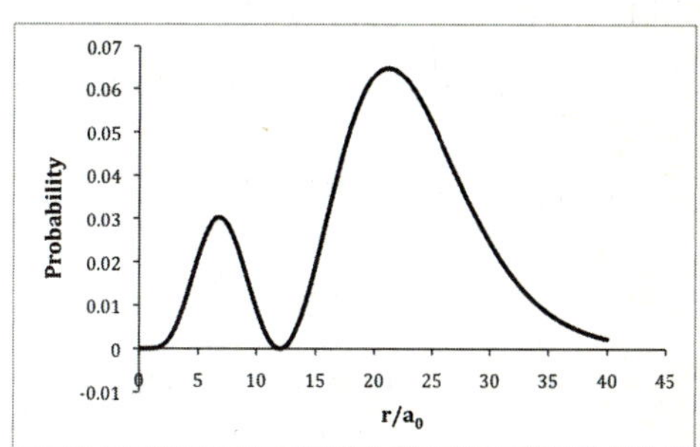

c.

5s

$4d_{z^2}$

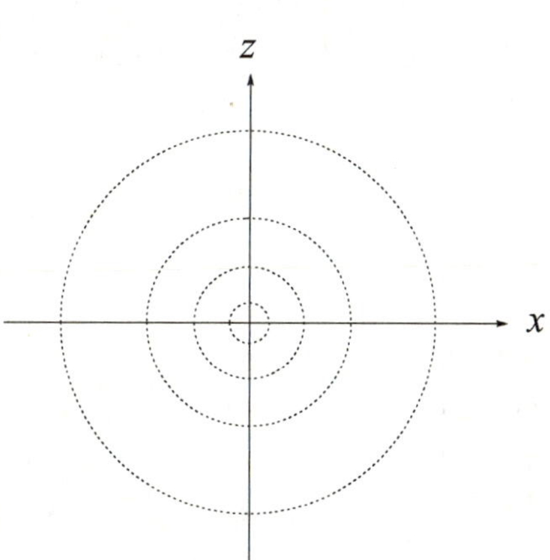

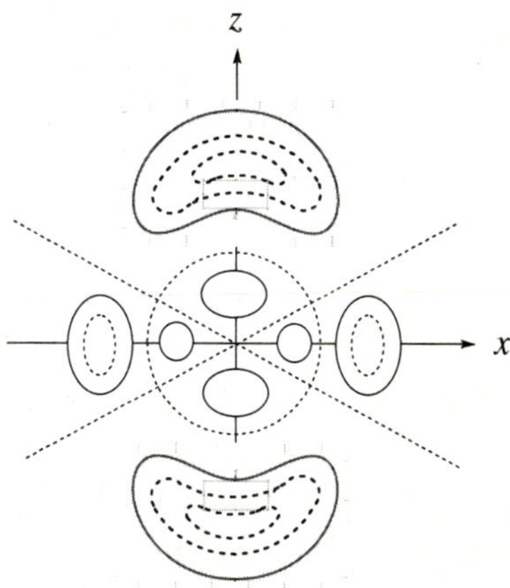

Dashed lines represent radial nodes. Electron density is inside the smallest node, between the nodes, and outside the largest node.

2.10 $4f_{z(x^2-y^2)}$ orbital

a. No radial nodes

b. 3 angular nodes

c. The angular nodes are solutions for $z(x^2-y^2) = 0$. These solutions are $z = 0$ (*xy* plane), and the planes where $x = y$ and $x = -y$, both perpendicular to the *xy* plane.

d. There are 8 lobes, 4 above and 4 below the *xy* plane. Down the *z* axis, this orbital looks like a $d_{x^2-y^2}$ orbital, but the node at the *xy* plane splits each lobe in two. For an image of this orbital, please see http://winter.group.shef.ac.uk/orbitron/ or another atomic orbital site on the Web.

2.11 An $5f_{xyz}$ orbital has the same general shape as the $4f_{z(x^2-y^2)}$, with the addition of a radial node and rotation so the lobes are between the *xy*, *xz*, and *yz* planes.

a. 1 radial node

b. 3 angular nodes

c. The angular nodes are solutions for $xyz = 0$. These solutions are the three planes where $x = 0$ (*yz* plane), $y = 0$ (*xz* plane), and $z = 0$ (*xy* plane).

d. The diagram is similar to that in problem 2.10, with a radial node added and rotated by 45° around the *z* axis. There are 8 lobes, one in each octant of the coordinate system, both inside and outside the radial node, for a total of 16 lobes. For an image of this orbital, please see http://winter.group.shef.ac.uk/orbitron/ or another atomic orbital site on the Web.

2.12 **a.** No radial nodes

b. 3 angular nodes

c. The angular nodes are solutions for $z(5z^2 - 3r^2) = 0$
Solutions are $z = 0$ (the *xy* plane) and $5z^2 - 3r^2 = 0$, or $5z^2 = 3r^2$
Because $r^2 = x^2 + y^2 + z^2$ we can write

$$5z^2 = 3(x^2 + y^2 + z^2)$$
$$2z^2 = 3x^2 + 3y^2$$
$$z^2 = 3/2\,(x^2 + y^2)$$

This is the equation for a (double) cone.

d. For an image of this orbital, please see http://winter.group.shef.ac.uk/orbitron/ or another atomic orbital site on the Web.

2.13 **a.**

	n	l	m_l
5*d*	5	2	–2,–1,0,1,2
4*f*	4	3	–3,–2,–1,0,1,2,3
7*g*	7	4	–4,–3,–2,–1,0,1,2,3,4

b.

	n	l	m_l	m_s
3d	3	2	–2	$\pm\frac{1}{2}$
	3	2	–1	$\pm\frac{1}{2}$
	3	2	0	$\pm\frac{1}{2}$
	3	2	1	$\pm\frac{1}{2}$
	3	2	2	$\pm\frac{1}{2}$

10 possible combinations.

c. For f orbitals ($l = 3$) possible values of m_l are –3, –2, –1, 0, 1, 2, and 3.

d. 2. (No more than 2 electrons can occupy any orbital!)

2.14 **a.** For a 5d electron, $l = 2$ and $n = 5$.

b. At most there can be ten 4d electrons, half of which can have $m_s = -\frac{1}{2}$.

c. f electrons have quantum number $l = 3$ and can have $m_l = -3, -2, -1, 0, 1, 2$, or 3.

d. For $l = 4$ (g electrons), m_l can be –4, –3, –2, –1, 0, 1, 2, 3, or 4.

2.15 **a.** The l quantum number limits the number of electrons. For $l = 3$, m_l can have seven values (–3 to +3) each defining an atomic orbital, and for each value of m_l the value of m_s can be $-\frac{1}{2}$ or $+\frac{1}{2}$; there can be two electrons in each orbital. At most, therefore, there can be 14 electrons with $n = 5$ and $l = 3$.

b. A 5d electron has $l = 2$, limiting the possible values of m_l to –2, –1, 0, 1, and 2.

c. p orbitals have $l = 1$ and occur for $n \geq 2$.

d. g orbitals have $l = 4$. There are 9 possible values of m_l and therefore 9 orbitals.

2.16 **a.**

↑ ↑	↑↓ __	
One exchange $E = \Pi_e$	One pair in same orbital $E = \Pi_c$	The first configuration is favored. This configuration is stabilized by Π_e (which is negative), and the second is destabilized by Π_c (which is positive).

b.

↑ ↑ ↑	and	↑↓ ↑ __	
3 exchanges (1–2, 1–3, 2–3) $E = 3\,\Pi_e$		1 exchange, 1 pair $E = \Pi_e + \Pi_c$	The first configuration is favored. It is stabilized by $2\Pi_e$ more than the second configuration, and the second configuration is also destabilized by Π_c.

2.17 **W:** ↑ ↑ ↓ ↓
2 exchanges (1–2, 3–4)
$E = 2\Pi_e$

X: ↑ ↑ ↑ ↓
3 exchanges (1–2, 1–3, 2–3)
$E = 3\Pi_e$, so this state has lower energy than **W** (because Π_e is negative).

2.18 Y: $\underline{\uparrow}\ \underline{\uparrow}\ \underline{\uparrow}\ \underline{\downarrow}\ \underline{\downarrow}$
4 exchanges (1–2, 1–3, 2–3, 4–5); $E = 4\Pi_e$

Z: $\underline{\uparrow}\ \underline{\uparrow}\ \underline{\uparrow}\ \underline{\uparrow}\ \underline{\downarrow}$
6 exchanges (1–2, 1–3, 1–4, 2–3, 2–4, 3–4)
$E = 6\Pi_e$, so this state has lower energy than **Y**. (because Π_e is negative).

2.19 **a.** Figure 2.12 and the associated text explain this. Electron-electron repulsion is minimized by placing each electron in a separate orbital when the levels are close enough to allow it. At Cr, the second 4*s* electron has an energy higher than the lower five 3*d* electrons, and therefore the configuration is $4s^1\ 3d^5$.

b. Ti is $4s^2\ 3d^2$, since both 4*s* electrons have energies below that of the 3*d* electrons at that *Z*. For ions, the 3*d* levels move down in energy, and are below the 4*s* levels for all transition metal M^{2+} ions, so Cr^{2+} is [Ar] $3d^4$.

2.20 **a.** V $1s^2\ 2s^2\ 2p^6\ 3s^2\ 3p^6\ 4s^2\ 3d^3$

b. Br $1s^2\ 2s^2\ 2p^6\ 3s^2\ 3p^6\ 4s^2\ 3d^{10}\ 4p^5$

c. Ru^{3+} $1s^2\ 2s^2\ 2p^6\ 3s^2\ 3p^6\ 3d^{10}\ 4s^2\ 4p^6\ 4d^5$

d. Hg^{2+} $1s^2\ 2s^2\ 2p^6\ 3s^2\ 3p^6\ 4s^2\ 3d^{10}\ 4p^6\ 5s^2\ 4d^{10}\ 5p^6\ 4f^{14}\ 5d^{10}$

e. Sb $1s^2\ 2s^2\ 2p^6\ 3s^2\ 3p^6\ 4s^2\ 3d^{10}\ 4p^6\ 5s^2\ 4d^{10}\ 5p^3$

2.21 **a.** Rb^- [Kr] $5s^2$

b. Pt^{2-} [Xe] $6s^2\ 4f^{14}\ 5d^{10}$

2.22 **a.** When a fluorine atom gains an electron, it achieves the same electron configuration of the noble gas Ne, with all subshells filled.

b. Zn^{2+} has the configuration [Ar] $3d^{10}$, with a filled 3*d* subshell.

c. This configuration [Kr] $5s^1\ 4d^5$ has each electron outside the Kr core in a separate orbital and has the maximum number of electrons with parallel spins (6). In Mo this configuration has a lower energy than [Kr] $5s^2\ 4d^4$. The greater stability of the s^1d^5 configuration is also found in the case of Cr (directly above Mo), but W (below Mo) has an s^2d^4 configuration, a reflection of the similar energies of the 6*s* and 5*d* orbitals.

2.23 **a.** Ag^+ has the configuration [Kr] $4d^{10}$, with a filled 4*d* subshell.

b. In Cm the $s^2d^1f^7$ configuration enables the last eight electrons each to occupy a separate orbital with parallel spin. This minimizes Π_c in comparison with the alternative configuration, in which one *f* orbital would be occupied by an electron pair.

c. The ion Sn^{2+} has the configuration [Kr] $5s^2\ 4d^{10}$, with both the 5*s* and 4*d* subshells filled. This is an example of the "inert pair" effect, in which heavier elements of groups 13-16 often form compounds in which the oxidation states of these elements are 2 less than the final digit in the group number (other examples include Tl^+ and Bi^{3+}). The reasons for this phenomenon are much more complex than the electron configurations of the ions; see N. N. Greenwood and A. Earnshaw, *Chemistry of the Elements*, 2nd ed., Butterworth-Heinemann, London, 1997, pp. 226–227 for useful comments.

2.24 **a.** Ti^{2+} has the configuration [Ar] $3d^2$; Ni^{2+} has the configuration [Ar] $3d^8$. In each case it should be noted that all the electrons outside the noble gas cores in these transition metal ions are *d* electrons, which occupy orbitals that are lower in energy than the 4*s* orbitals (see Figure 2.12(b) and the associated discussion).

b. The preferred configuration of Mn^{2+} is [Ar] $3d^5$. As in the examples in part a, the 3*d* orbitals of Mn^{2+} are lower in energy than the 4*s*. In addition, the configuration minimizes electron–electron repulsions (because each *d* electron is in a separate orbital) and maximizes the stabilizing effect of electrons with parallel spins (maximum Π_e).

2.25 **a.**

	Z $(1s^2)$ $(2s^2\,2p^6)$ $(3s^2\,3p^n)$	Z^*	r
	1 0.85 0.35		
P	$15 - (2 \times 1 + 8 \times 0.85 + 4 \times 0.35) =$	4.8	106 pm
S	$16 - (2 \times 1 + 8 \times 0.85 + 5 \times 0.35) =$	5.45	102 pm
Cl	$17 - (2 \times 1 + 8 \times 0.85 + 6 \times 0.35) =$	6.1	99 pm
Ar	$18 - (2 \times 1 + 8 \times 0.85 + 7 \times 0.35) =$	6.75	98 pm

The size of the atoms decreases slightly as Z increases, even though the number of electrons in the atom increases, because Z^* increases and draws the electrons closer. Ar has the strongest attraction between the nucleus and the 3*p* electron, and the smallest radius.

b.

	Z $(1s^2)$ $(2s^2\,2p^6)$	Z^*	r
	0.85 0.35		
O^{2-}	$8 - (2 \times 0.85 + 7 \times 0.35) =$	3.85	126 pm
F^-	$9 - (2 \times 0.85 + 7 \times 0.35) =$	4.85	119 pm
Na^+	$11 - (2 \times 0.85 + 7 \times 0.35) =$	5.85	116 pm
Mg^{2+}	$12 - (2 \times 0.85 + 7 \times 0.35) =$	6.85	86 pm

These values increase directly with Z, and parallel the decrease in ionic size. Increasing nuclear charge results in decreasing size for these isoelectronic ions, although the change between F^- and Na^+ is smaller than might be expected.

c.

Cu	$(1s^2)\,(2s^2\,2p^6)\;(3s^2\,3p^6)\;(3d^{10})\,(4s^1)$		
4*s*	$S = 2 + 8 + (8 \times 0.85) + (10 \times 0.85) =$	25.3	$Z^* = 29 - 25.3 = 3.7$
3*d*	$S = 2 + 8 + (8 \times 1.00) + (9 \times 0.35) =$	21.15	$Z^* = 29 - 21.15 = 7.85$

The 3*d* electron has a much larger effective nuclear charge and is held more tightly; the 4*s* electron is therefore the first removed on ionization.

d. $(1s^2)\ (2s^2\ 2p^6)\ (3s^2\ 3p^6)\ (3d^{10})\ (4s^2\ 4p^6)\ (4d^{10})\ (4f^n)$

Ce $S = 2 + 8 + 8 + 10 + 8 + 10 = 46$

[Xe] $6s^2\ 4f^1\ 5d^1$ — $4f^1$ — $Z^* = 58 - 46 = 12$

Pr $S = 2 + 8 + 8 + 10 + 8 + 10 + (2 \times 0.35) = 46.7$

[Xe] $6s^2\ 4f^3$ — $4f^3$ — $Z^* = 59 - 46.7 = 12.3$

Nd $S = 2 + 8 + 8 + 10 + 8 + 10 + (3 \times 0.35) = 47.05$

[Xe] $6s^2\ 4f^4$ — $4f^4$ — $Z^* = 60 - 47.05 = 12.95$

The outermost electrons experience an increasing Z^*, and are therefore drawn in to slightly closer distances with increasing Z and Z^*.

2.26

		Sc:	Ti:
3d electron:			
1	Electron configuration	$(1s^2)(2s^2, 2p^6)(3s^2, 3p^6)(3d^1)(4s^2)$	$(1s^2)(2s^2, 2p^6)(3s^2, 3p^6)(3d^2)(4s^2)$
4a	Contribution of other d electrons	None [only one electron in $(3d^1)$]	One other $3d$ electron: Contribution to $S = 1 \times 0.35 = 0.35$
4b	Contribution of electrons to left of $(3d^n)$	$18 \times 1.00 = 18.00$	$18 \times 1.00 = 18.00$
	Total S	18.00	18.35
	Z^*	$21 - 18.00 = 3.00$	$22 - 18.35 = 3.65$
4s electron:			
1	Electron configuration	$(1s^2)(2s^2, 2p^6)(3s^2, 3p^6)(3d^1)(4s^2)$	$(1s^2)(2s^2, 2p^6)(3s^2, 3p^6)(3d^2)(4s^2)$
3a	Contribution of other $(4s^2)$ electron	Contribution to $S = 0.35$	Contribution to $S = 0.35$
3b	Contribution of $(3s^2, 3p^6)(3d^n)$ electrons	$9 \times 0.85 = 7.65$	$10 \times 0.85 = 8.50$
3c	Contribution of other electrons	$10 \times 1.00 = 10.00$	$10 \times 1.00 = 10.00$
	Total S	$0.35 + 7.65 + 10.00 = 18.00$	$0.35 + 8.50 + 10.00 = 18.85$
	Z^*	$21 - 18.00 = 3.00$	$22 - 18.85 = 3.15$

In Sc Slater's rules give the same value for the effective nuclear charge Z^* for the $3d$ and $4s$ electrons, consistent with the very similar energies of these orbitals. In Ti the effective nuclear charge is slightly less for $4s$ than for $3d$, by 0.50. This difference in the energies of $4s$ and $3d$ increases across the row of transition metals; by nickel, as calculated in the example in section 2.2.4, the $4s$ orbital has an effective nuclear charge 3.50 less than the $3d$ orbitals. This is consistent with the experimental observation that transition metal cations have electron configurations in which there are no valence s electrons; the $3d$ orbitals are lower in energy than the $4s$ orbital, so d orbitals are the ones that are occupied. A similar phenomenon is observed for the second and third row transition metals.

2.27

N	$Z^* = 7 - (2 \times 0.85 + 4 \times 0.35) = 3.9$	IE = 1.402 MJ/mol	$r = 75$ pm
P	$Z^* = 15 - (2 \times 1 + 8 \times 0.85 + 4 \times 0.35) = 4.80$	IE = 1.012 MJ/mol	$r = 106$ pm
As	$Z^* = 33 - (2 \times 1 + 8 \times 1 + 18 \times 0.85 + 4 \times 0.35) = 6.3$	IE = 0.947 MJ/mol	$r = 120$ pm

The effect of shielding alone is not sufficient to explain the changes in ionization energy. The other major factor is distance between the electron and the nucleus. Wulfsberg (*Principles of Descriptive Inorganic Chemistry*, Brooks/Cole, 1998) suggests that Z^*/r^2 correlates better, but As is still out of order:

	radius(pm)	Z^*/r^2
N	75	6.93×10^{-4}
P	106	4.27×10^{-4}
As	120	4.375×10^{-4}

2.28 Zr through Pd have the electron configurations shown below:

40	Zr	$5s^2\ 4d^2$
41	Nb	$5s^1\ 4d^4$
42	Mo	$5s^1\ 4d^5$
43	Tc	$5s^2\ 4d^5$
44	Ru	$5s^1\ 4d^7$
45	Rh	$5s^1\ 4d^8$
46	Pd	$4d^{10}$

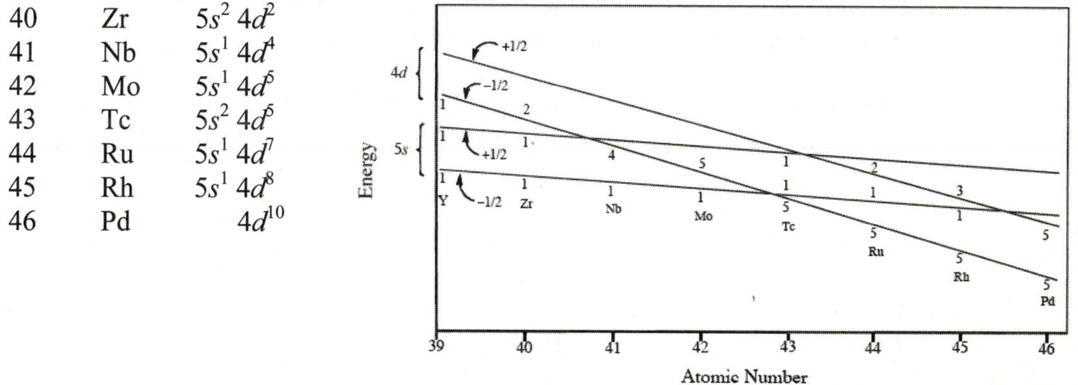

The lower d line crosses the upper s line between elements 40 and 41, the upper d line crosses the upper s line between 43 and 44, and the upper d line crosses the lower s line between 45 and 46. This graph fits the experimental configurations well.

2.29

	Li	>	Na	>	K	>	Rb
IE	5.39		5.14		4.34		4.18
Z^*	1.30		2.20		2.20		2.20
r(pm, covalent)	123		154		203		216
Z^*/r^2	8.59		9.28		5.34		4.72 (all $\times 10^{-5}$ pm^{-2})
r(pm, ionic)	90		116		152		166
Z^*/r^2	16.0		16.3		9.52		7.98 (all $\times 10^{-5}$ pm^{-2})

The Z^*/r^2 function explains the order, except for lithium, which seems to require more energy for removal of the electron than predicted. Apparently the very small size and small number of electrons on lithium result in the electron being held more tightly than in the other alkali metals. The Z^*/r^2 function also predicts larger differences between the IE values than are observed.

2.30 C^+ $(1s^2)$ $(2s^2\ 2p^1)$

$$Z^* = 6 - (2 \times 0.85 + 2 \times 0.35) = 3.6$$

B $(1s^2)$ $(2s^2\ 2p^1)$

$$Z^* = 5 - (2 \times 0.85 + 2 \times 0.35) = 2.6$$

The energies change by a factor of approximately three, but Z^* changes only by 38%. Based on the data in Table 2.8 and the relative ionic and covalent sizes in problem 29, C^+ has a radius of about 58 pm, B a radius of about 83 pm. Z^*/r^2 values are then 10.7×10^{-4} and 3.8×10^{-4}, a ratio of 2.8. The ionization energies have a ratio of 2.99. This function of size and effective charge explains this pair quite well.

2.31 Because the second ionization energies involve removing an electron from an ion with a single positive charge, the electron configurations of the cations must be considered:

He^+	$1s^1$	N^+	$1s^2\,2s^2\,2p^2$
Li^+	$1s^2$	O^+	$1s^2\,2s^2\,2p^3$
Be^+	$1s^2\,2s^1$	F^+	$1s^2\,2s^2\,2p^4$
B^+	$1s^2\,2s^2$	Ne^+	$1s^2\,2s^2\,2p^5$
C^+	$1s^2\,2s^2\,2p^1$		

The peaks and valleys now match the peaks and valleys for electron configurations matching those of the atoms in the ionization energy graph in Figure 2.13. For example, because there is a minimum in the ionization energy for Li (electron configuration $1s^2\,2s^1$), there should also be a minimum in the graph of second ionization energy for Be^+, which also has the configuration $1s^2\,2s^1$. The maximum in the ionization energy for Be (configuration $1s^2\,2s^2$) would be matched by a maximum in the second ionization graph for B^+, which has the same configuration as Be. Overall, the maxima and minima in the two graphs are:

Maximum or Minimum	Configuration	1st IE	2nd IE (kJ/mol)	
Maximum	$1s^2$	He	Li^+	(7298)
Minimum	$1s^2\,2s^1$	Li	Be^+	(1757)
Maximum	$1s^2\,2s^2$	Be	B^+	(2427)
Minimum	$1s^2\,2s^2\,2p^1$	B	C^+	(2353)
	$1s^2\,2s^2\,2p^2$	C	N^+	(2856)
Maximum	$1s^2\,2s^2\,2p^3$	N	O^+	(3388)
Minimum	$1s^2\,2s^2\,2p^4$	O	F^+	(3374)
	$1s^2\,2s^2\,2p^5$	F	Ne^+	(3952)

2.32 **a.** Fe (7.87 eV) > Ru (7.37 eV) They have the same Z^* (6.25), but Ru is larger, so Z^*/r^2 is smaller than for Fe.

b. P (10.486 eV) > S (10.36 eV) Z^* is smaller for P (4.8) than for S (5.45), but S has one electron paired in the 3*p* level, which increases its energy and makes it easier to remove.

c. Br (11.814 eV) > K (4.341) Z^* for K is 2.2; for Br it is 7.6, a very large difference. K is also nearly twice as large as Br. Both factors contribute to the difference in IE.

d. N (14.534) > C (11.260) Increasing Z^* (3.9 for N, 3.25 for C) and decreasing size (75 pm for N, 77 pm for C) both work in the same direction.

e. Cd (8.993) > In (5.786) Indium starts a new 5*p* subshell, so the last electron is easily removed in spite of a larger Z^* (2.30 for In, 1.65 for Cd).

f. The smaller F has a higher ionization energy (17.422) than Cl (12.967). An electron is more easily removed from the 3*p* subshell of Cl than from the smaller 2*p* subshell of F.

2.33 a. S (2.077 eV) has a smaller EA than Cl (3.617 eV) because Cl has a larger Z^* (6.1 vs. 5.45) and a slightly smaller radius. Both increase the attractive power for an electron.

b. I (3.059) has a smaller EA than Br (3.365) because it is larger, with the same Z^* (7.6).

c. B (8.298) has a smaller IE than Be (9.322) because it is starting a new *p* subshell.

d. S (10.360) has a smaller IE than P (10.486) because S^- is losing one of a pair of *p* electrons.

2.34 a. The maximum at Mg comes at a completed subshell ($3s^2$) and the minimum at Al is at the first electron of a new subshell (3*p*), increasing the energy and making removal of an electron easier. The maximum at P is at a half-filled subshell ($3p^3$) and the minimum at S is at the fourth 3*p* electron, which must pair with one of the others; this also raises the energy of the electron and makes its removal easier.

b. The reasons for the minima and maxima in the electron affinity graph are the same as in the ionization energy graph. The maxima and minima are shifted by one in the two graphs because the reactants in the process defining electron affinity have a negative charge, one more electron than a neutral atom. For example, minima occur for ionization energy at Al ([Ne]$3s^23p^1$ $\rightarrow$) and for electron affinity at Mg ([Ne]$3s^23p^1$ $\rightarrow$); Al and Mg^- have identical electron configurations.

2.35 The Bohr equation (p.19) predicts that the energy levels of 1-electron systems should be proportional to Z^2. $Z^2 = 4$ for He^+ and 9 for Li^{2+}, so ratios of the ionization energies to that of H are 4:1 and 9:1.

2.36 In both the transition metals and the lanthanides (problem 25d), the gradual change in Z^* pulls the outer electrons closer. The increase in Z^* is 0.65 for each unit increase in atomic number, so the increase in attraction is relatively small, and the change in radius must also be small.

2.37 a. $Se^{2-} > Br^- > Rb^+ > Sr^{2+}$ These ions are isoelectronic, so the sizes are directly dependent on the nuclear charges. For example, Sr^{2+} has the greatest nuclear charge and is the smallest ion.

b. $Y^{3+} > Zr^{4+} > Nb^{5+}$ These ions are also isoelectronic, so the increasing nuclear charge results in decreasing size.

c. $Co^{4+} < Co^{3+} < Co^{2+} < Co$ The smaller number of electrons with constant nuclear charge results in a smaller size.

2.38 **a.** F^- has the largest radius. All three choices have 10 electrons, and F^- has the fewest protons (9) to attract these electrons.

b. Te^{2-} has the largest volume. In general, atomic and ionic sizes increase going down a column of main group elements as the number of shells increased, with electrons closer to the nucleus shielding outer electrons from the full effect of the nuclear charge.

c. Mg, with a filled 3*s* subshell, has the highest ionization energy. The situation is similar to that in the second period, where Be has a higher ionization energy than Li and B.

d. Fe^{3+} has the greatest surplus of positive charge (26 protons, 23 neutrons), making it the smallest and the most difficult from which to remove an electron.

e. Of the anions O^-, F^-, and Ne^-, F^- has the configuration of the noble gas Ne, from which it is more difficult to remove an electron than from the other ions; fluorine, therefore, has the highest electron affinity.

2.39 **a.** V is the smallest; the effective nuclear charge on the outer electrons increases across the row of transition metals.

b. These species are isoelectronic; each has 18 electrons. S^{2-}, which has the fewest protons, is therefore the largest.

c. Because ionization energy decreases with increasing size for the alkali metals, Cs has the lowest value.

d. The electron affinity is the greatest for the smallest halogens (stronger attraction between nucleus and outermost electrons), so Cl has the highest value.

e. The Cu^{2+} ion has the greatest surplus of protons and is therefore the smallest and the most difficult from which to remove an electron.

2.40 **a.** 7 orbitals

b. See images.

c. The number of radial nodes increases as *n* increases: 4*f* orbitals have no radial nodes 5*f* have one radial node, and so forth.

2.41 **a.** 9 orbitals

b. See images.

c. As in 2.40.c., with no radial nodes for 5*g*, one for 6*g*, etc.

CHAPTER 3: SIMPLE BONDING THEORY

3.1 **a.** Structures a and b are more likely than c because the negative formal charge is on the electronegative S. In c, the highly electronegative N has a positive charge.

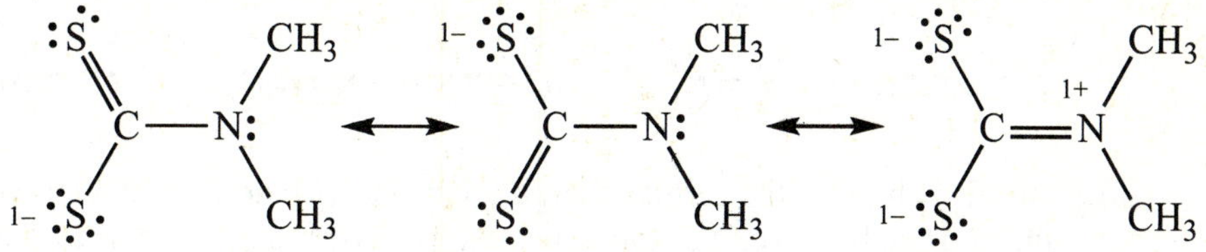

b. The same structures fit $[OSCN(CH_3)_2]^-$. The structure with a 1– formal charge on O is most likely, since O is the most electronegative atom in the ion.

3.2 **a.**

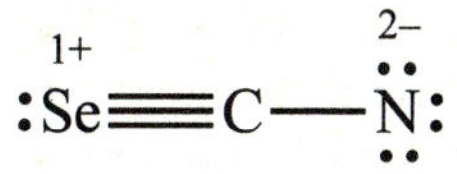

$:Se\equiv C—\ddot{N}:$ (1+ on Se, 2– on N) The formal charges are large, but match electronegativity.

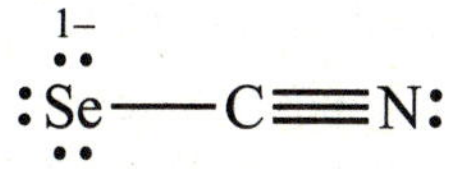

$:\ddot{Se}—C\equiv N:$ (1– on Se) Negative formal charge of 1– on Se, a low electronegativity atom.

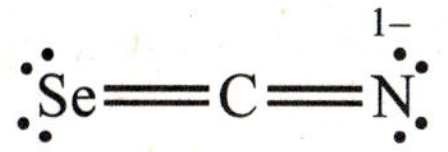

$Se=C=N$ (1– on N) Negative formal charge on N, the most electronegative atom. Best resonance structure of the three.

b. b is better than a, because the formal charge is on the more electronegative O.

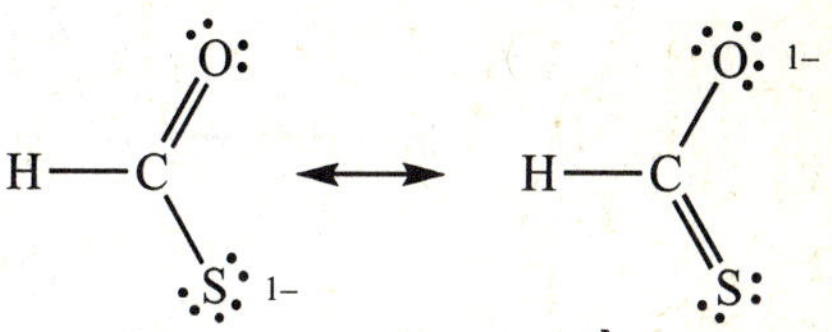

c. a and b are better than c, because one of the formal charges is on the more electronegative O.

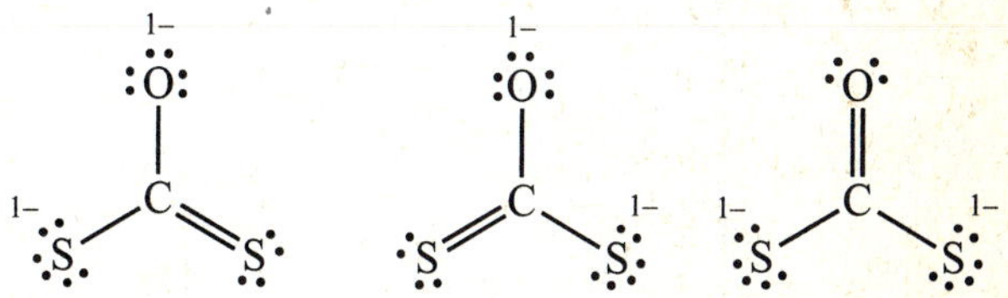

3.3 NSO^-: a has 2– formal charge on N, 1+ on S. Large formal charges, not very likely. b has a 1– formal charges on N and O, 1+ on S, and is a better structure.

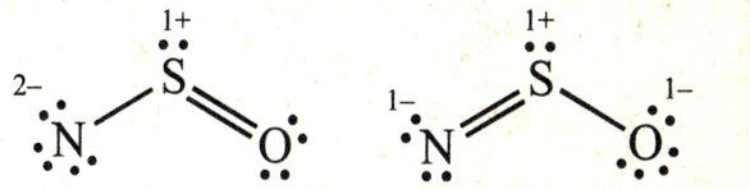

SNO^-: a has a 1– formal charge on S. Not very likely, doesn't match electronegativity (negative formal charge is not on most electronegative atoms). b has 1– formal charge on O, and is a better structure.

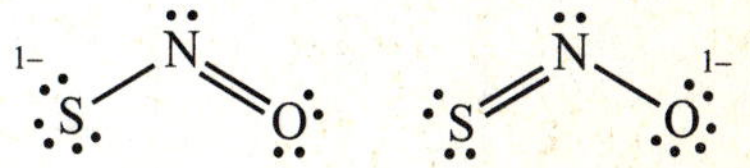

Overall, the $S=N–O^-$ structure is better based on formal charges, since it has only a negative charge on O, the most electronegative atom in the ion.

3.4

	I	II	III
A	O=N=C=N (N 1+, N 1−)	O=N=N=C (N 1+, N 1+, C 2−)	N=O=C=N (N 1−, O 2+, N 1−)
B	O=N−C≡N	O=N−N≡C (N 1+, C 1−)	N=O−C≡N (N 1−, O 1+)
C	O≡N−C=N (O 1+, N 1+, C 1−, N 1−)	O≡N−N=C (O 1+, N 1+, C 2−)	N≡O−C=N (O 2+, C 1−, N 1−)

Structure IB is best by the formal charge criterion, with no formal charges, and is expected to be the most stable. None of the structures II or III are as good; they have unlikely charges (by electronegativity arguments) or large charges.

3.5 :N≡N−O: (N 1+, O 1−) ⟷ N=N=O (N 1−, N 1+) ⟷ :N−N≡O: (N 2−, N 1+, O 1+)

The first resonance structure, which places the negative formal charge on the most electronegative atom, provides a slightly better representation than the second structure, which has its negative formal charge on the slightly less electronegative nitrogen. Experimental measurements show that the nitrogen–nitrogen distance (112.6 pm) in N_2O is slightly closer to the triple bond distance (109.8 pm) in N_2 than to the double bond distances found in other nitrogen compounds, and thermochemical data are also consistent with the first structure providing the best representation. The third resonance structure, with greater overall magnitudes of formal charges, is the poorest representation.

3.6 H−O−N(=O)−O: 1− ⟷ H−O−N(−O: 1−)=O

3.7

Molecule	*Atom*	*Group Number*	*Unshared Electrons*	$2\left(\frac{\chi_A}{\chi_A+\chi_B}\right)$	*Number of Bonds*	*Formal Charge*
C≡O	C	4	2	$2\left(\frac{2.544}{2.544+3.61}\right)=0.83$	3	−0.49
	O	6	2	$2\left(\frac{3.61}{2.544+3.61}\right)=1.17$	3	0.49
$N{=}O^-$	N	5	4	$2\left(\frac{3.066}{3.066+3.61}\right)=0.92$	2	−0.84
	O	6	4	$2\left(\frac{3.61}{3.066+3.61}\right)=1.08$	2	−0.16
H–F	H	1	0	$2\left(\frac{2.300}{2.300+4.193}\right)=0.71$	1	0.29
	F	7	6	$2\left(\frac{4.193}{2.300+4.193}\right)=1.29$	1	−0.29

Surprisingly, CO is more polar than FH, and NO^- is intermediate, with C and N the negative atoms in CO and NO^-.

3.8 **a.** $SeCl_4$ requires 10 electrons around Se. The lone pair of electrons in an equatorial position of a trigonal bipyramid distorts the shape by bending the axial chlorines back.

b. I_3^- requires 10 electrons around the central I and is linear.

c. $PSCl_3$ is nearly tetrahedral. The multiple bonding in the P–S bond compresses the Cl—P—Cl angles to 101.8°, significantly less than the tetrahedral angle.

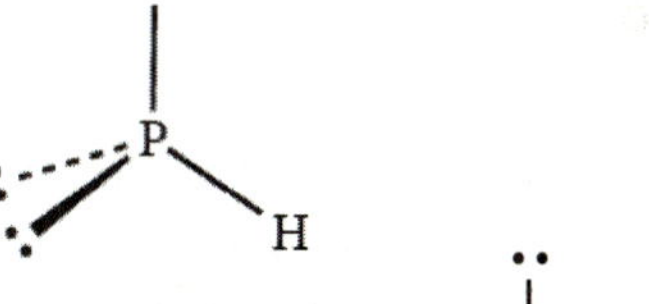

d. IF_4^- has 12 electrons around I and has a square planar shape.

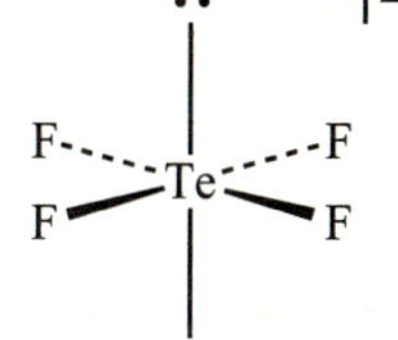

e. PH_2^- has a bent structure, with two lone pairs.

f. TeF_4^{2-} has 12 electrons around Te, with a square planar shape.

g. N_3^- is linear, with two double bonds in its best resonance structure.

h. $SeOCl_4$ has a distorted trigonal bipyramidal shape with the extra repulsion of the double bond placing oxygen in an equatorial position.

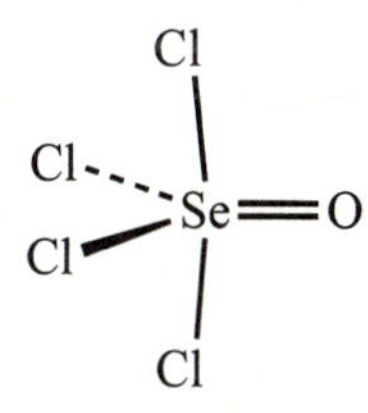

i. PH_4^+ is tetrahedral.

3.9 **a.** ICl_2^- has 10 electrons around I and is linear.

b. H_3PO_3 has a distorted tetrahedral shape.

c. BH_4^- is tetrahedral.

d. $POCl_3$ is a distorted tetrahedron. The Cl—P—Cl angle is compressed to 103.3° as a result of the phosphorus–oxygen double bond.

e. IO_4^- is tetrahedral, with significant double bonding; all bonds are equivalent.

f. $IO(OH)_5$ has the oxygens arranged octahedrally, with hydrogens on five of the six oxygens.

g. $SOCl_2$ is trigonal pyramidal, with one lone pair and some double bond character in the S–O bond.

h. $ClOF_4^-$ is a square pyramid. The double bonded O and the lone pair occupy opposite positions.

i. The F—Xe—F angle is nearly linear (174.7°), with the two oxygens and a lone pair in a trigonal planar configuration. Formal charges favor double bond character in the Xe–O bonds. The O—Xe—O angle is narrowed to 105.7° by *lp–bp* repulsion.

3.10 a. SOF_6 is nearly octahedral around the S.

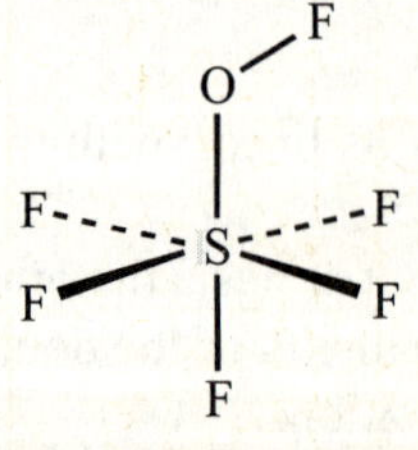

b. POF_3 has a distorted tetrahedral shape, with F—P—F angles of 101.3°.

c. ClO_2 is an odd electron molecule, with a bent shape, partial double bond character, and an angle of 117.5°.

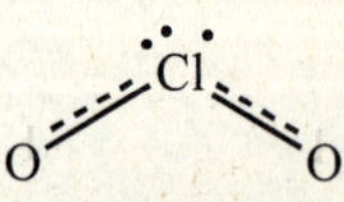

d. NO_2 is another odd electron molecule, bent, with partial double bond character and an angle of 134.25°. This is larger than the angle of ClO_2 because there is only one odd electron on N, rather than the one pair and single electron of ClO_2.

e. $S_2O_4^{2-}$ has SO_2 units with an angle of about 30° between their planes, in an eclipsed conformation.

f. N_2H_4 has a trigonal pyramidal shape at each N, and a *gauche* conformation. There is one lone pair on each N.

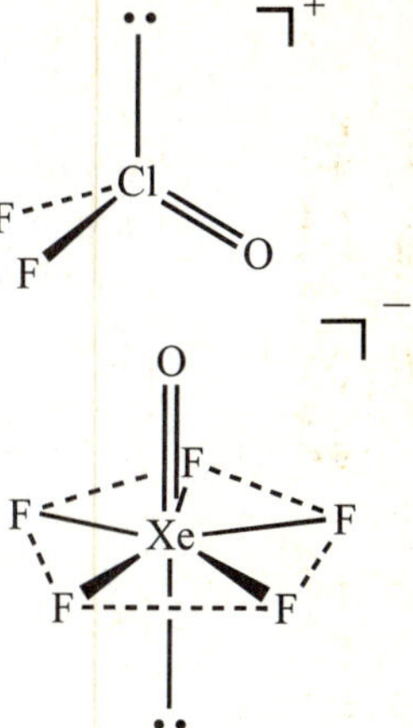

g. $ClOF_2^+$ is a distorted trigonal pyramid with one lone pair and double bond character in the chlorine–oxygen bond.

h. CS_2, like CO_2, is linear with double bonds. S=C=S

i. The structure of $XeOF_5^-$ is based on a pentagonal bipyramid, with a lone pair and the oxygen atom in axial positions. See K. O. Christe et al., *Inorg. Chem.*, **1995**, *34*, 1868 for evidence in support of this structure.

3.11 All the halate ions are trigonal pyramids; as the central atom increases in size, the bonding pairs are farther from the center, and the lone pair forces a smaller angle. The decreasing electronegativity Cl > Br > I of the central atom also allows the electrons to be pulled farther out, reducing the *bp-bp* repulsion.

3.12 **a.** AsH_3 should have the smallest angle, since it has the largest central atom. This minimizes the bond pair–bond pair repulsions and allows a smaller angle. Arsenic is also the least electronegative central atom, allowing the electrons to be drawn out farther and lowering the repulsions further. Actual angles: $AsH_3 = 91.8°$, $PH_3 = 93.8°$, $NH_3 = 106.6°$.

b. Cl is larger than F, and F is more electronegative and should pull the electrons farther from the S, so the F—S—F angle should be smaller in OSF_2. This is consistent with the experimental data: the F—S—F angle in OSF_2 is 92.3° and the Cl—S—Cl angle in $OSCl_2$ is 96.2°.

c. NO_2^- has rather variable angles (115° and 132°) in different salts. The sodium salt (115.4°) has a slightly smaller angle than O_3 (116.8°). The N–O electronegativity difference should pull electrons away from N, reducing the *bp-bp* repulsion and the angle.

d. BrO_3^- (104°) has a slightly smaller angle than ClO_3^- (107°), since it has a larger central atom. In addition, the greater electronegativity of Cl holds the electrons closer and increases *bp-bp* repulsion.

3.13 **a.** N_3^- is linear, with two double bonds. O_3 is bent (see solution to problem 12c), with one double bond and a lone pair on the central O caused by the extra pair of electrons.

b. Adding an electron to O_3 decreases the angle, as the odd electron spends part of its time on the central O, making two positions for electron repulsion. The decrease in angle is small, however, with angles of 113.0 to 114.6 pm reported for alkali metal ozonides (see W. Klein, K. Armbruster, and M. Jansen, *Chem. Commun.*, **1998**, 707) in comparison with 116.8° for ozone.

3.14 Cl–O–Cl 110.9° H$_3$C–O–CH$_3$ 111.8° H$_3$Si–O–SiH$_3$ 144.1°

As the groups attached to oxygen become less electronegative, the oxygen atom is better able to attract shared electrons to itself, increasing the electron–electron repulsions and increasing the bond angle. In the case of $O(SiH_3)_2$, the very large increase in bond angle over $O(CH_3)_2$ suggests that the size of the SiH_3 group also has a significant effect on the bond angle.

3.15 C_3O_2 has the linear structure O=C=C=C=O, with zero formal charges.

N_5^+ with the same electronic structure has formal charges of 1–, 1+, 1+, 1+, 1–, unlikely because three positive charges are adjacent to each other. Changing to N=N=N–N≡N results in formal charges of 1–, 1+, 0, 1+, 0, a more reasonable result with an approximately trigonal angle in the middle. With triple bonds on each end, the formal charges are 0, 1+, 1–, 1+, 0 and a tetrahedral angle. Some contribution from this would reduce the bond angle.

OCNCO$^+$ can have the structure O≡C–N–C≡O, with formal charges of 1+, 0, 1–, 0, 1+ and two lone pairs on the central N. This would result in an even smaller angle in the middle, but has positive formal charges on O, the most electronegative atom. O=C=N–C≡O has a formal charge of 1+ on the final O. Resonance would reduce that formal charge, making this structure and a trigonal angle more likely. The Seppelt reference also mentions two lone pairs on N and cites "the markedly higher electronegativity of the nitrogen atom with respect to the central atom in C_3O_2, which leads to a higher localization of electron density in the sense of a nonbonding electron pair." Therefore, the bond angles should be OCCCO > OCNCO$^+$ > N_5^+. Literature values are 180°, 130.7°, and 108.3 to 112.3° (calculated), respectively.

3.16 **a.**

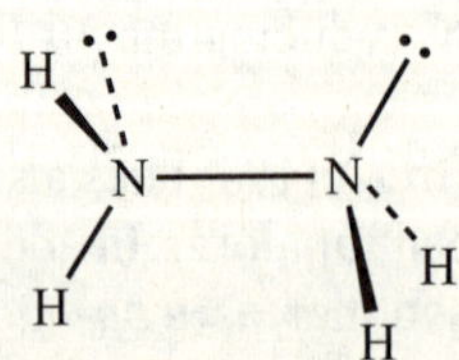

In ethylene, carbon has *p* orbitals not involved in sigma bonding. These orbitals interact to form a pi bond between the carbons, resulting in planar geometry. (Sigma and pi bonding are discussed further in Chapter 5.) In hydrazine each nitrogen has a steric number of 4, and there is sigma bonding only; the steric number of 4 requires a 3-dimensional structure.

b.

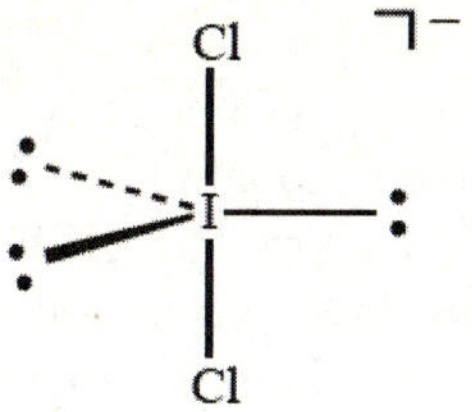

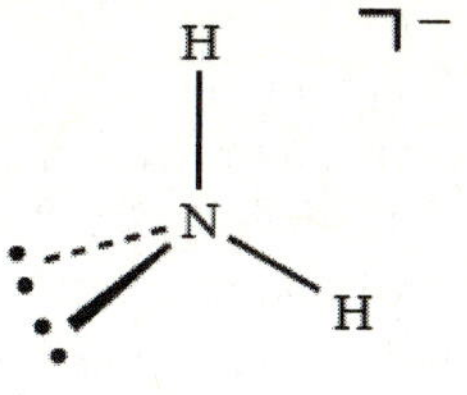

In ICl_2^- the iodine has a steric number of 5, with three lone pairs in equatorial positions; the consequence is a linear structure, with Cl atoms occupying axial positions. In NH_2^- the two lone pairs require a bent arrangement.

c. Resonance structures of cyanate and fulminate are shown in Figures 3.4 and 3.5. The fulminate ion has no resonance structures that have formal charges as low as in structures A and B for cyanate. The guideline that resonance structures having low formal charges tend to describe relatively stable structures is followed here. $Hg(CNO)_2$, in which the anion has higher formal charges in its resonance structures, is the explosive compound.

3.17 a. PCl_5 has 10 electrons around P, using 3*d* orbitals in addition to the usual 3*s* and 3*p*. N is too small to allow this structure. In addition, N would require use of the 3*s*, 3*p*, or 3*d* orbitals, but they are too high in energy to be used effectively.

b. Similar arguments apply, with O too small and lacking in accessible *d* orbitals.

3.18 a. The lone pairs in both molecules are equatorial, the position that minimizes 90° interactions between lone pairs and bonding pairs.

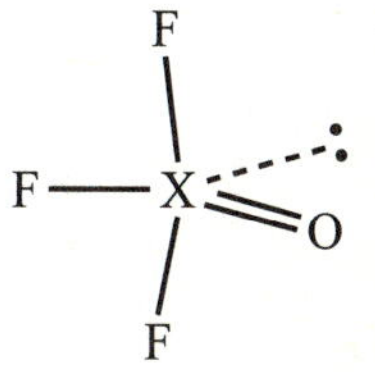

b. In $BrOF_3$ the less electronegative central atom allows electrons in the bonds to be pulled toward the F and O atoms to a greater extent, reducing repulsions near the central atom and enabling a smaller bond angle. In $BrOF_3$, the F_{eq}–Br–O angle is approximately 4.5° smaller than the comparable angle in $ClOF_3$.

3.19 IF_3^{2-} has three lone pairs and three bonds. Overall, this ion is predicted to be T-shaped, with bond angles slightly less than 90°.

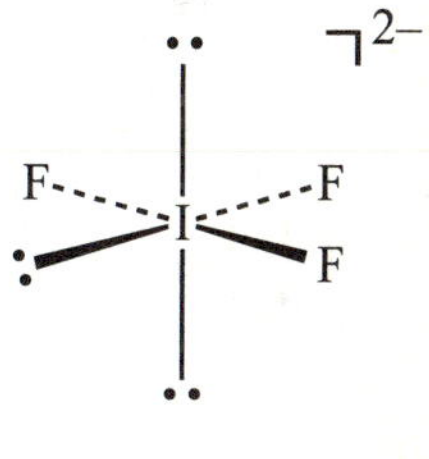

3.20 a. There are three possibilities:

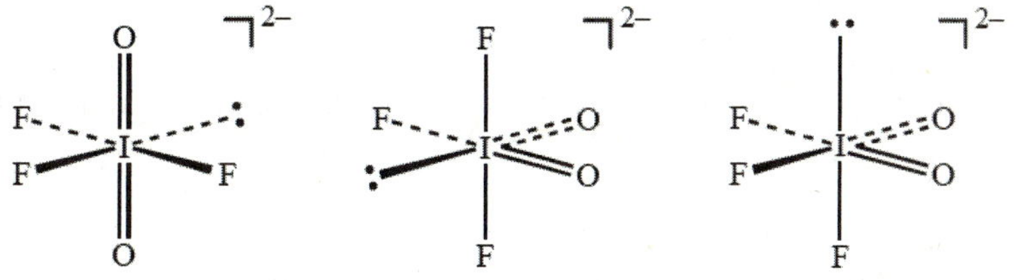

b. The third structure, with the lone pair and double bonds in a facial arrangement, is least likely because it would have the greatest degree of electron–electron repulsions involving these regions of high electron concentrations.

The second structure, which has fewer 90° lone pair–double bond repulsions than the first structure, is expected to be the most likely. Experimental data are most consistent with this structure.

c. One possibility: $XeO_2F_3^-$

3.21 $I(CF_3)Cl_2$ is roughly T-shaped, with the two Cl atoms opposite each other and the CF_3 group and two lone pairs in the trigonal plane. The experimental Cl—I—C angles are 88.7° and 82.9°, smaller than the 90° expected if there were no extra repulsion from the lone pairs. Repulsion between the lone pairs and the larger CF_3 group put them in the trigonal plane, where there is more room.

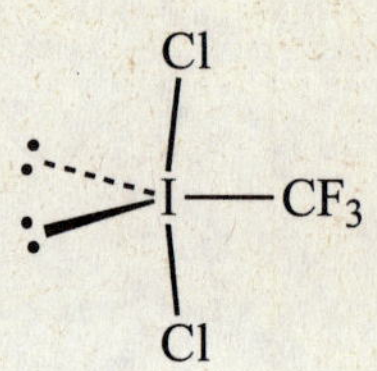

3.22 **a.** CF_3 has a greater attraction for electrons than CH_3, so the P in $PF_2(CF_3)_3$ is more positive than the P in $PF_2(CH_3)_3$. This draws the F atoms in slightly, so the P—F bonds are shorter in $PF_2(CF_3)_3$ (160.1 pm vs. 168.5 pm).

b. Al—O—Al could have an angle near 109°, like water, or could have double bonds in both directions and a nearly linear structure. In fact, the angle is about 140°. The single-bonded picture is more probable; the high electronegativity of O compared to Al draws the bonding pairs closer, opening up the bond angle. A Lewis structure with zero formal charges on all atoms can be drawn for this molecule with four electrons on each Al.

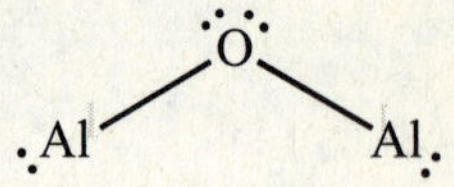

c. CAl_4 is tetrahedral. Again, a Lewis structure with zero formal charges can be drawn with four electrons on each Al.

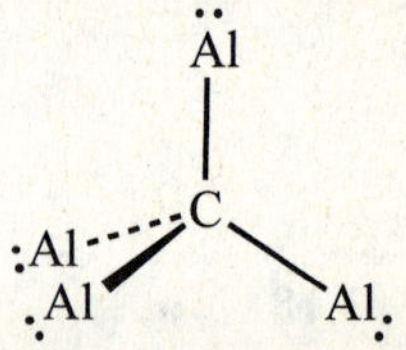

3.23

Octahedral	Distorted
$SeCl_6^{2-}$	SeF_6^{2-}
$TeCl_6^{2-}$	IF_6^-
ClF_6^-	

The distorted structures have the smallest outer atoms in comparison with the size of the central atom. In these cases there apparently is room for a lone pair to occupy a position that can lead to distortion. In the octahedral cases there may be too much crowding to allow a lone pair to distort the shape.

3.24

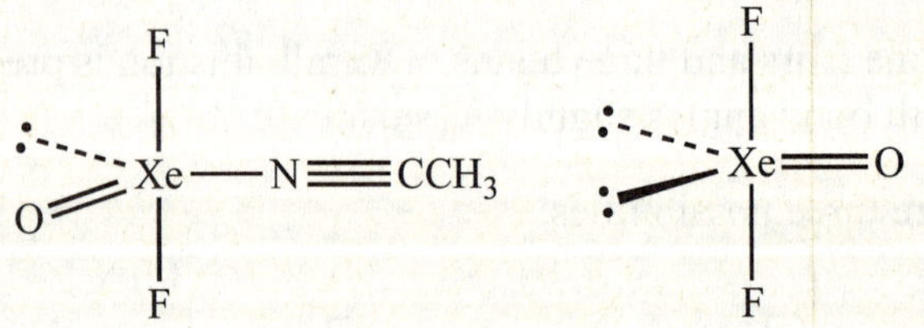

In $F_2OXeN{\equiv}CCH_3$ the nitrogen–xenon bond is weak; see the reference for details on bond distances and angles.

3.25 **a.** O is more electronegative than N and can draw the electrons more strongly away from the S. The more positive S in $OSCl_2$ consequently attracts bonding pairs in S–Cl bonds closer to sulfur, increasing *bp-bp* repulsions and increasing the Cl—S—Cl angle (96.2° in $OSCl_2$, 93.3° in $NSCl_2^-$).

b. Because the sulfur in $OSCl_2$ attracts the S–Cl bonding pairs more strongly, these bonds are shorter: 207.6 pm in $OSCl_2$, 242.3 pm in $NSCl_2^-$.

3.26 The larger, less electronegative Br atoms are equatorial.

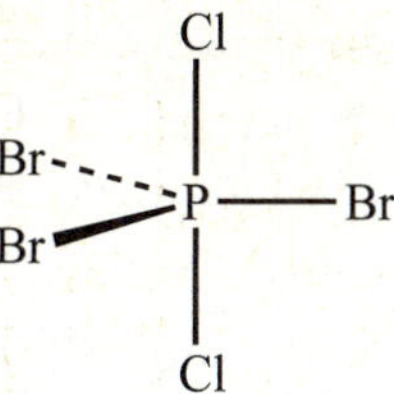

3.27 **a.** In $PCl_3(CF_3)_2$ the highly electronegative CF_3 groups occupy axial positions.

b. The axial positions in $SbCl_5$ experience greater repulsions by bonding pairs, leading to longer Sb–Cl (axial) bonds (223.8 pm) than Sb–Cl (equatorial) bonds (227.7 pm).

3.28 PF_4^+ has the bond angle expected for a tetrahedron, 109.5°. In PF_3O the multiple bond to oxygen results in distortion away from the oxygen, leading to a smaller F–P–F angle. By the LCP approach, the F···F distances should be approximately the same in these two structures. They are similar: 238 pm in PF_4^+ and 236 pm in PF_3O.

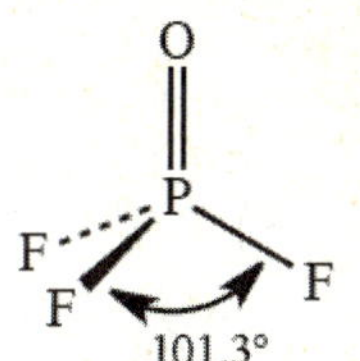

3.29 As more (less electronegative) CH_3 groups are added, there is greater concentration of electrons near P, and greater electron–electron repulsion leads to longer axial P–F bonds.

Reported P–F distances:	$PF_4(CH_3)$	$PF_3(CH_3)_2$	$PF_2(CH_3)_3$
	161 pm	164 pm	168 pm

3.30 Bond angles and distances:

	Steric Number	C—F (pm)	FCF angle (°)	F—F (pm)
$F_2C{=}CF_2$	3	133.6	109.2	218
F_2CO	3	131.9	107.6	216
CF_4	4	131.9	109.5	216
F_3CO^-	4	139.2	101.3	215

The differences between these molecules are subtle. The LCP model views the F ligands as hard objects, tightly packed around the central C in these examples. In this approach, the F···F distance remains nearly constant while the central atom moves to minimize repulsions.

3.31 The calculation is similar to the example shown in Section 3.2.4.

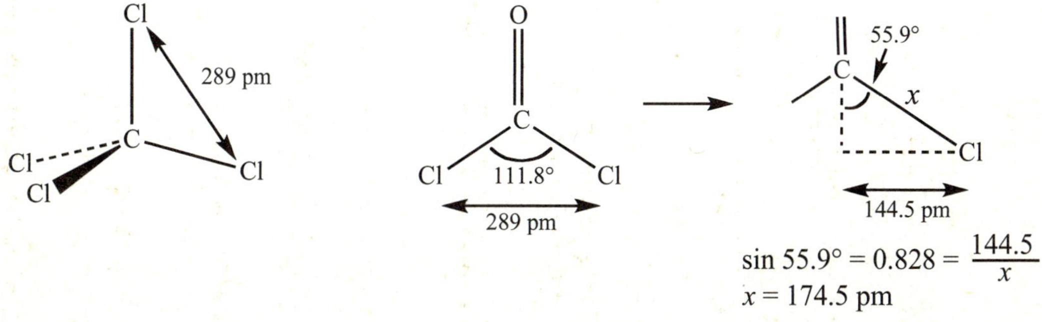

$$\sin 55.9° = 0.828 = \frac{144.5}{x}$$

$x = 174.5$ pm

3.32 By the LCP approach, from the structures of HOH and FOF the hydrogen radius would be 76 pm (half of the H···H distance) and the fluorine radius (half of the F···F distance) would be 110 pm. Because the LCP model describes nonbonded outer atoms as being separated by the sums of their radii, as if they were touching spheres, the H···F distance in HOF would therefore be the sum of the ligand radii, 76 + 110 = 186 pm, in comparison with the actual H···F distance of 183 pm. If the covalent O–H and O–F bonds in HOF are similar to the matching distances in HOH and FOF, the H–O–F angle must be smaller than the other angles because of the H···F distance.

An alternative explanation considers the polarity in HOF. Because of the high electronegativity of fluorine, the F atom in HOF acquires a partial negative charge, which is attracted to the relatively positive H atom. By this approach, electrostatic attraction between H and F reduces the bond angle in HOF, giving it the smallest angle of the three compounds.

3.33 The electronegativity differences are given in parentheses:

a. C–N N is negative (0.522)

b. N–O O is negative (0.544)

c. C–I C is negative (0.185)

d. O–Cl O is negative (0.741)

e. P–Br Br is negative (0.432)

f. S–Cl Cl is negative (0.280)

The overall order of polarity is O–Cl > N–O > C–N > P–Br > S–Cl > C–I.

3.34 **a.** $VOCl_3$ has a distorted tetrahedral shape, with Cl—V—Cl angles of 111°, and Cl—V—O angles of 108°.

b. PCl_3 has a trigonal pyramidal shape with Cl—P—Cl angles of 100.4°.

c. SOF_4 has a distorted trigonal bipyramidal shape. The axial fluorine atoms are nearly linear with the S atom; the equatorial F—S—F angle is 100°.

d. SO_3 is trigonal with equal bond angles of 120°.

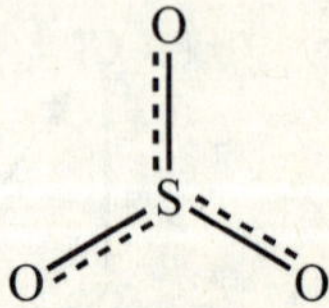

e. ICl_3 would be expected to have two axial lone pairs, causing distortion to reduce the Cl (axial)–I–Cl (equatorial) angles to < 90°. However, reaction of I_2 with Cl_2 yields dimeric I_2Cl_6, which readily dissociates into ICl and Cl_2.

f. SF_6 is a regular octahedron.

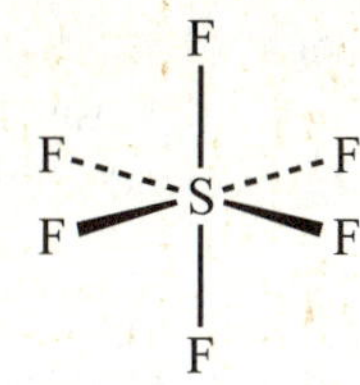

g. IF_7 is a rare example of pentagonal bipyramidal geometry.

h. The structure of XeO_2F_4 is based on an octahedron, with oxygens in *trans* positions because of multiple bonding.

i. CF_2Cl_2, like methane, is tetrahedral.

j. P_4O_6 is described in the problem. Each P has one lone pair, each O has two.

3.35 **a.** PH_3 has a smaller bond angle than NH_3, about 93°. The larger central atom reduces the repulsion between the bonding pairs.

b. H_2Se has a structure like water, with a bond angle near 90°. The larger central atom increases the distance between the S–H bonding pairs and reduces their repulsion, resulting in a smaller angle than in water.

c. SeF_4 has a lone pair at one of the equatorial positions of a trigonal bipyramid, and bond angles of about 110° (equatorial) and 169° (axial). Teeter-totter shape.

d. PF_5 has a trigonal bipyramidal structure.

e. IF_5 is square pyramidal, with slight distortion away from the lone pair.

f. XeO_3 has a trigonal pyramidal shape, similar to NH_3, but with Xe–O double bonds.

g. BF_2Cl is trigonal planar, with ∠FBCl larger than ∠FBF.

h. $SnCl_2$ has a bond angle of 95° in the vapor phase, smaller than the trigonal angle. As a solid, it forms polymeric chains with bridging chlorines and bond angles near 80°.

i. KrF_2 is linear: F—Kr—F. VSEPR predicts three lone pairs on krypton in equatorial positions, with the fluorine atoms in axial positions.

j. $IO_2F_5^{2-}$ has a steric number of 7 on iodine, with oxygen atoms occupying axial positions.

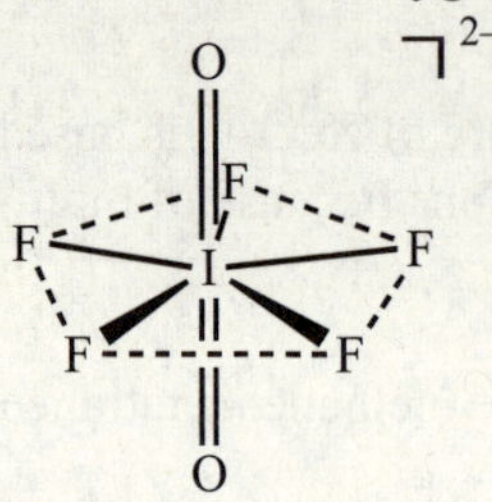

3.36 Polar: $VOCl_3$, PCl_3, SOF_4, ICl_3, CF_2Cl_2

3.37 Polar: PH_3, H_2Se, SeF_4, IF_5, XeO_3, BF_2Cl, $SnCl_2$

3.38 **a.** The H–O bond of methanol is more polar than the H–S bond of methyl mercaptan. As a result, hydrogen bonding holds the molecules together and requires more energy for vaporization. The larger molecular weight of methyl mercaptan has a similar effect, but the hydrogen bonding in methanol has a stronger influence.

b. CO and N_2 have nearly identical molecular weights, but the polarity of CO leads to dipole–dipole attractions that help hold CO molecules together in the solid and liquid states.

c. The *ortho* isomer of hydroxybenzoic acid can form *intra*molecular hydrogen bonds, while the *meta* and *para* isomers tend to form dimers and larger aggregates in their hydrogen bonding. As a result of their better ability to form hydrogen bonds between molecules (*inter*molecular hydrogen bonds), the *meta* and *para* isomers have higher melting points (*ortho*, 159°; *meta*, 201.3°; *para*, 214–215°).

Intramolecular hydrogen bond

d. The London (dispersion) forces between atoms increase with the number of electrons, so the noble gases with larger *Z* have larger interatomic forces and higher boiling points.

e. Acetic acid can form hydrogen-bonded dimers in the gas phase, so the total number of particles in the gas is half the number expected by using the ideal gas law.

f. Acetone has a negative carbonyl oxygen; chloroform has a positive hydrogen, due to the electronegative character of the chlorines. As a result, there is a stronger attraction between the different kinds of molecules than between molecules of the same kind, and a resulting lower vapor pressure. (This is an unusual case of hydrogen bonding, with no H–N, H–O, or H–F bond involved.)

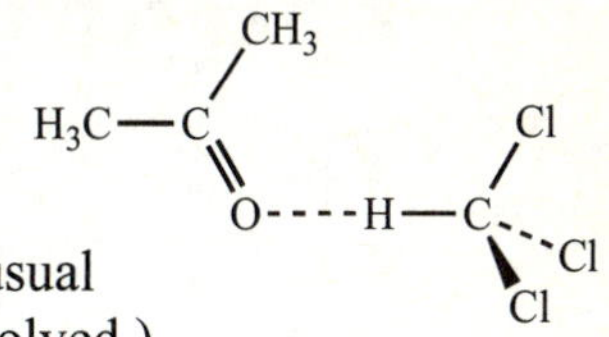

CH_3 $H_3C—C$ O- - - -$H—C$ Cl Cl Cl

g. CO has about 76 kJ contribution to its bond energy because of the electronegativity difference between C and O; attraction between the slightly positive and negative ends strengthens the bonding. Although this is not a complete explanation, it covers most of the difference between CO and N_2. In spite of its high bond energy, N_2 is thought by some to have some repulsion in its sigma bonding because of the short bond distance.

CHAPTER 4: SYMMETRY AND GROUP THEORY

4.1 **a.** Ethane in the staggered conformation has 2 C_3 axes (the C–C line), 3 perpendicular C_2 axes bisecting the C–C line, in the plane of the two C's and the H's on opposite sides of the two C's. No σ_h, $3\sigma_d$, i, S_6. Overall, a $\boldsymbol{D_{3d}}$ molecule.

b. Ethane in eclipsed conformation has two C_3 axes (the C–C line), three perpendicular C_2 axes bisecting the C–C line, in the plane of the two C's and the H's on the same side of the two C's. Mirror planes include σ_h and $3\sigma_d$. Overall, a $\boldsymbol{D_{3h}}$ molecule.

c. Chloroethane in the staggered conformation has only one mirror plane, through both C's, the Cl, and the opposite H on the other C. Overall, a $\boldsymbol{C_s}$ molecule.

d. 1,2-dichloroethane in the *trans* conformation has a C_2 axis perpendicular to the C–C bond and perpendicular to the plane of both Cl's and both C's, a σ_h plane through both Cl's and both C's, and an inversion center. Overall, a $\boldsymbol{C_{2h}}$ molecule.

4.2 **a.** Ethylene is a planar molecule, with C_2 axes through the C's and perpendicular to the C–C bond both in the plane of the molecule and perpendicular to it. It also has a σ_h plane and two σ_d planes (arbitrarily assigned). Overall, a $\boldsymbol{D_{2h}}$ molecule.

b. Chloroethylene is also a planar molecule, with the only symmetry element the mirror plane of the molecule. Therefore, a $\boldsymbol{C_s}$ molecule.

c. 1,1-dichloroethylene has a C_2 axis coincident with the C–C bond, and two mirror planes, one the plane of the molecule and one perpendicular to the plane of the molecule through both C's. Overall, a $\boldsymbol{C_{2v}}$ molecule.

cis-1,2-dichloroethylene has a C_2 axis perpendicular to the C–C bond and in the plane of the molecule, two mirror planes (one in the plane of the molecule and one perpendicular to the plane of the molecule and perpendicular to the C–C bond). Overall a $\boldsymbol{C_{2v}}$ molecule.

trans-1,2-dichloroethylene has a C_2 axis perpendicular to the C–C bond and perpendicular to the plane of the molecule, a mirror plane in the plane of the molecule, and an inversion center. Overall, a $\boldsymbol{C_{2h}}$ molecule.

4.3 **a.** Acetylene has a C_∞ axis through all four atoms, an infinite number of perpendicular C_2 axes, a σ_h plane, and an infinite number of σ_d planes through all four atoms. Overall, a $\boldsymbol{D_{\infty h}}$ molecule.

b. Fluoroacetylene has only the C_∞ axis through all four atoms and an infinite number of mirror planes, also through all four atoms. Overall a $\boldsymbol{C_{\infty v}}$ molecule.

c. Methylacetylene has a C_3 axis through the carbons and three σ_v planes, each including one hydrogen and all three C's. Overall, a $\boldsymbol{C_{3v}}$ molecule.

d. 3-Chloropropene (assuming a rigid molecule, no rotation around the C–C bond) has no rotation axes and only one mirror plane through Cl and all three C atoms. Overall, a $\boldsymbol{C_s}$ molecule.

e. Phenylacetylene (again assuming no internal rotation) has a C_2 axis down the long axis of the molecule and two mirror planes, one the plane of the benzene ring and the other perpendicular to it. The point group is $\boldsymbol{C_{2v}}$.

4.4 a. Napthalene has three perpendicular C_2 axes, and a horizontal mirror plane (regardless of which C_2 is taken as the principal axis), making it a $\boldsymbol{D_{2h}}$ molecule.

b. 1,8-dichloronaphthalene has only one C_2 axis, the C–C bond joining the two rings, and two mirror planes, making it a $\boldsymbol{C_{2v}}$ molecule.

c. 1,5-dichloronaphthalene has one C_2 axis perpendicular to the plane of the molecule, a horizontal mirror plane, and an inversion center; overall, $\boldsymbol{C_{2h}}$.

d. 1,2-dichloronaphthalene has only the mirror plane of the molecule, and is a $\boldsymbol{C_s}$ molecule.

4.5 a. 1,1′-dichloroferrocene has a C_2 axis parallel to the rings, perpendicular to the Cl–Fe–Cl σ_h mirror plane. It also has an inversion center; overall, $\boldsymbol{C_{2h}}$.

b. Dibenzenechromium has collinear C_6, C_3, and C_2 axes perpendicular to the rings, six perpendicular C_2 axes and a σ_h plane, making it a $\boldsymbol{D_{6h}}$ molecule. It also has three σ_v and three σ_d planes, S_3 and S_6 axes, and an inversion center.

c. Benzenebiphenylchromium has a mirror plane through the Cr and the biphenyl bridge bond. It has no other symmetry elements, so it is a $\boldsymbol{C_s}$ molecule.

d. H_3O^+ has the same symmetry as NH_3: a C_3 axis, and three σ_v planes for a $\boldsymbol{C_{3v}}$ molecule.

e. O_2F_2 has a C_2 axis perpendicular to the O–O bond and perpendicular to a line connecting the fluorines. With no other symmetry elements, it is a $\boldsymbol{C_2}$ molecule.

f. Formaldehyde has a C_2 axis collinear with the C=O bond, a mirror plane including all the atoms, and another perpendicular to the first and including the C and O atoms. Overall, $\boldsymbol{C_{2v}}$.

g. S_8 has C_4 and C_2 axes perpendicular to the average plane of the ring, four C_2 axes through opposite bonds, and four mirror planes perpendicular to the ring, each including two S atoms. Overall, $\boldsymbol{D_{4d}}$.

h. Borazine has a C_3 axis perpendicular to the plane of the ring, three perpendicular C_2 axes, and a horizontal mirror plane. Overall, $\boldsymbol{D_{3h}}$.

i. Tris(oxalato)chromate(III) has a C_3 axis and three perpendicular C_2 axes, each splitting a C–C bond and passing through the Cr. Overall, $\boldsymbol{D_3}$.

j. A tennis ball has three perpendicular C_2 axes (one through the narrow portions of each segment, the others through the seams) and two mirror planes including the first rotation axis. Overall, $\boldsymbol{D_{2d}}$.

4.6 **a.** Cyclohexane in the chair conformation has a C_3 axis perpendicular to the average plane of the ring, three perpendicular C_2 axes between the carbons, and three σ_v planes, each including the C_3 axis and one of the C_2 axes. Overall, $\boldsymbol{D_{3d}}$.

b. Tetrachloroallene has three perpendicular C_2 axes, one collinear with the double bonds, and the other two at 45° to the Cl—C—Cl planes. It also has two σ_v planes, one defined by each of the Cl–C–Cl groups. Overall, $\boldsymbol{D_{2d}}$. (Note that the ends of tetrachlorallene are staggered.)

c. The sulfate ion is tetrahedral. $\boldsymbol{T_d}$.

d. Most snowflakes have hexagonal symmetry (Figure 4.2), and have collinear C_6, C_3 and C_2 axes, six perpendicular C_2 axes, and a horizontal mirror plane. Overall, $\boldsymbol{D_{6h}}$. (For high quality images of snowflakes, including some that have different shapes, see K. G. Libbrecht, *Snowflakes*, Voyageur Press, Minneapolis, MN, 2008.)

e. Diborane has three perpendicular C_2 axes and three perpendicular mirror planes. $\boldsymbol{D_{2h}}$.

f. 1,3,5-tribromobenzene has a C_3 axis perpendicular to the plane of the ring, three perpendicular C_2 axes, and a horizontal mirror plane. $\boldsymbol{D_{3h}}$.

1,2,3-tribromobenzene has a C_2 axis through the middle Br and two perpendicular mirror planes which include this axis. $\boldsymbol{C_{2v}}$.

1,2,4-tribromobenzene has only the plane of the ring as a mirror plane. $\boldsymbol{C_s}$.

g. A tetrahedron inscribed in a cube has $\boldsymbol{T_d}$ symmetry (see Figure 4.6).

h. The left and right ends of B_3H_8 are staggered with respect to each other. There is a C_2 axis through the borons. In addition, there are two planes of symmetry, each containing four H atoms, and two C_2 axes between these planes and perpendicular to the original C_2. The point group is $\boldsymbol{D_{2d}}$.

4.7 **a.** A sheet of typing paper has three perpendicular C_2 axes and three perpendicular mirror planes. Overall, $\boldsymbol{D_{2h}}$.

b. An Erlenmeyer flask has an infinite-fold rotation axis and an infinite number of σ_v planes. $\boldsymbol{C_{\infty v}}$.

c. A screw has no symmetry operations other than the identity, for a $\boldsymbol{C_1}$ classification.

d. The number 96 (with the correct type font) has a C_2 axis perpendicular to the plane of the paper, making it $\boldsymbol{C_{2h}}$.

e. Your choice—the list is too long to attempt to answer it here.

f. A pair of eyeglasses has only a vertical mirror plane. $\boldsymbol{C_s}$.

g. A 5-pointed star has a C_5 axis, five perpendicular C_2 axes, one horizontal and five vertical mirror planes. $\boldsymbol{D_{5h}}$.

h. A fork has only a mirror plane. $\boldsymbol{C_s}$.

i. Captain Hook has no symmetry operation other than the identity. $\boldsymbol{C_1}$.

j. A metal washer has a C_∞ axis, an infinite number of perpendicular C_2 axes, an infinite number of σ_v mirror planes, and a horizontal mirror plane. Overall, $\boldsymbol{D_{\infty h}}$.

4.8

a.	D_{2h}	**f.**	C_3
b.	D_{4h} (note the four knobs)	**g.**	C_{2h}
c.	C_s	**h.**	C_{8v}
d.	C_{2v}	**i.**	$D_{\infty h}$
e.	C_{6v}	**j.**	C_{3v}

4.9

a.	D_{3h}	**f.**	C_s (note holes)
b.	D_{4h}	**g.**	$C_{\infty v}$ (ignoring decorations)
c.	C_s	**h.**	C_{3v}
d.	C_3	**i.**	$D_{\infty h}$
e.	C_{2v}	**j.**	C_1

4.10 Hands (of identical twins): C_2 Baseball: D_{2d} Atomium: C_{3v}

Eiffel Tower: C_{4v} Dominoes: 6×6: C_{2v} 3×3: C_2 5×4: C_s

Bicycle wheel: The wheel shown has 32 spokes. The point group assignment depends on how the pairs of spokes (attached to both the front and back of the hub) connect with the rim. If the pairs alternate with respect to their side of attachment, the point group is D_{8d}. Other arrangements are possible, and different ways in which the spokes cross can affect the point group assignment; observing an actual bicycle wheel is recommended. (If the crooked valve is included, there is no symmetry, and the point group is a much less interesting C_1.)

4.11 a. Problem 3.34:

a. $VOCl_3$: C_{3v}	**e.** ICl_3: C_{2v}	**h.** XeO_2F_4: D_{4h}
b. PCl_3: C_{3v}	**f.** SF_6: O_h	**i.** CF_2Cl_2: C_{2v}
c. SOF_4: C_{2v}	**g.** IF_7: D_{5h}	**j.** P_4O_6: T_d
d. SO_3: D_{3h}		

b. Problem 3.35:

a. PH_3: C_{3v} | **e.** IF_5: C_{4v} | **h.** $SnCl_2$: C_{2v}

b. H_2Se: C_{2v} | **f.** XeO_3: C_{3v} | **i.** KrF_2: $D_{\infty h}$

c. SeF_4: C_{2v} | **g.** BF_2Cl: C_{2v} | **j.** $IO_2F_5^{2-}$: D_{5h}

d. PF_5: D_{3h}

4.12 a. Figure 3.8:

a. CO_2: $D_{\infty h}$ | **d.** PCl_5: D_{3h} | **f.** IF_7: D_{5h}

b. SO_3: D_{3h} | **e.** SF_6: O_h | **g.** TaF_8^{3-}: D_{4d}

c. CH_4: T_d

b. Figure 3.15:

a. CO_2: $D_{\infty h}$ | **e.** SNF_3: C_{3v} | **i.** SOF_4: C_{2v}

b. COF_2: C_{2v} | **f.** SO_2Cl_2: C_{2v} | **j.** ClO_2F_3: C_{2v}

c. NO_2^-: C_{2v} | **g.** XeO_3: C_{3v} | **k.** XeO_3F_2: D_{3h}

d. SO_3: D_{3h} | **h.** SO_4^{2-}: T_d | **l.** IOF_5: C_{4v}

4.13 a. p_x has $C_{\infty v}$ symmetry. (Ignoring the difference in sign between the two lobes, the point group would be $D_{\infty h}$.)

b. d_{xy} has D_{2h} symmetry. (Ignoring the signs, the point group would be D_{4h}.)

c. $d_{x^2-y^2}$ has D_{2h} symmetry. (Ignoring the signs, the point group would be D_{4h}.)

d. d_{z^2} has $D_{\infty h}$ symmetry.

e. f_{xyz} has T_d symmetry.

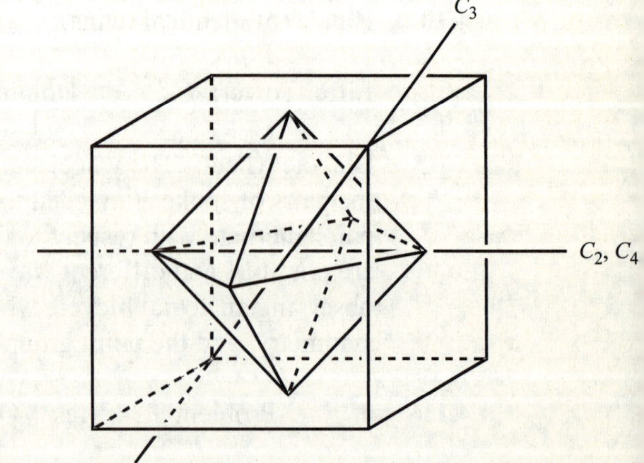

4.14 a. The superimposed octahedron and cube show the matching symmetry elements.

The descriptions below are for the elements of a cube; each element also applies to the octahedron.

E Every object has an identity operation.

8 C_3 Diagonals through opposite corners of the cube are C_3 axes.

6 C_2 Lines bisecting opposite edges are C_2 axes.

6 C_4 Lines through the centers of opposite faces are C_4 axes. Although there are only three such lines, there are six axes, counting the C_4^3 operations.

$3C_2$ $(=C_4^2)$ The lines through the centers of opposite faces are C_4 axes as well as C_2 axes.

i The center of the cube is the inversion center.

$6S_4$ The C_4 axes are also S_4 axes.

$8S_6$ The C_3 axes are also S_6 axes.

$3\sigma_h$ These mirror planes are parallel to the faces of the cube.

$6\sigma_d$ These mirror planes are through two opposite edges.

b. O_h

c. O

4.15 **a.** on-deck circle $D_{\infty h}$

b. batter's box D_{2h}

c. cap C_s

d. bat $C_{\infty v}$

e. home plate C_{2v}

f. baseball D_{2d} (see Figure 4.1)

g. pitcher C_1

4.16

a. D_{2h}

b. C_{2v}

c. C_{2v}

d. D_{4h}

e. C_{5h}

f. C_{2v}

g. D_{2h}

h. D_{4h}

i. C_{2h}

4.17 SNF_3

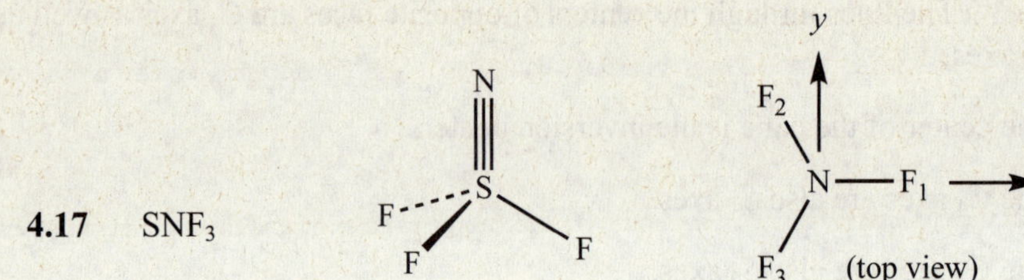

Symmetry Operations

F_2, N—F_1, F_3	F_1, N—F_3, F_2	F_3, N—F_1, F_2
after E	after C_3	after $\sigma_v(xz)$

Matrix Representations (reducible)

$$E: \begin{bmatrix} 1 & 0 & 0 \\ 0 & 1 & 0 \\ 0 & 0 & 1 \end{bmatrix} \quad C_3: \begin{bmatrix} \cos\frac{2\pi}{3} & -\sin\frac{2\pi}{3} & 0 \\ \sin\frac{2\pi}{3} & \cos\frac{2\pi}{3} & 0 \\ 0 & 0 & 1 \end{bmatrix} = \begin{bmatrix} -\frac{1}{2} & -\frac{\sqrt{3}}{2} & 0 \\ \frac{\sqrt{3}}{2} & -\frac{1}{2} & 0 \\ 0 & 0 & 1 \end{bmatrix} \quad \sigma_v(xz): \begin{bmatrix} 1 & 0 & 0 \\ 0 & -1 & 0 \\ 0 & 0 & 1 \end{bmatrix}$$

Characters of Matrix Representations

3	0	1

Block Diagonalized Matrices

$$\begin{bmatrix} \begin{bmatrix} 1 & 0 \\ 0 & 1 \end{bmatrix} & \begin{matrix} 0 \\ 0 \end{matrix} \\ \begin{matrix} 0 & 0 \end{matrix} & [1] \end{bmatrix} \qquad \begin{bmatrix} \begin{bmatrix} -\frac{1}{2} & -\frac{\sqrt{3}}{2} \\ \frac{\sqrt{3}}{2} & -\frac{1}{2} \end{bmatrix} & \begin{matrix} 0 \\ 0 \end{matrix} \\ \begin{matrix} 0 & 0 \end{matrix} & [1] \end{bmatrix} \qquad \begin{bmatrix} \begin{bmatrix} 1 & 0 \\ 0 & -1 \end{bmatrix} & \begin{matrix} 0 \\ 0 \end{matrix} \\ \begin{matrix} 0 & 0 \end{matrix} & [1] \end{bmatrix}$$

Irreducible Representations

E	$2C_3$	$3\sigma_v$	Coordinates used
2	–1	0	(x, y)
1	1	1	z

Character Table

C_{3v}	E	$2C_3$	$3\sigma_v$	Matching Functions	
A_1	1	1	1	z	x^2+y^2, z^2
A_2	1	1	–1	R_z	
E	2	–1	0	$(x, y), (R_x, R_y)$	$(x^2-y^2, xy)(xz, yz)$

4.18 **a.** C_{2h} molecules have E, C_2, i, and σ_h operations.

Cl H
C=C
H Cl

b. E: $\begin{matrix} 1 & 0 & 0 \\ 0 & 1 & 0 \\ 0 & 0 & 1 \end{matrix}$ C_2: $\begin{matrix} -1 & 0 & 0 \\ 0 & -1 & 0 \\ 0 & 0 & 1 \end{matrix}$ i: $\begin{matrix} -1 & 0 & 0 \\ 0 & -1 & 0 \\ 0 & 0 & -1 \end{matrix}$ σ_v: $\begin{matrix} 1 & 0 & 0 \\ 0 & 1 & 0 \\ 0 & 0 & -1 \end{matrix}$

c. These matrices can be block diagonalized into three 1 × 1 matrices, with the representations shown in the table.

	$\chi(E)$	$\chi(C_2)$	$\chi(i)$	$\chi(\sigma_h)$	
B_u	1	–1	–1	1	from the x and y coefficients
A_u	1	1	–1	–1	from the z coefficients

The total is $\Gamma = 2B_u + A_u$.

d. Multiplying B_u and A_u:
$1 \times 1 + (-1) \times 1 + (-1) \times (-1) + 1 \times (-1) = 0$, proving they are orthogonal

4.19 **a.** D_{2h} molecules have E, $C_2(z)$, $C_2(y)$, $C_2(x)$, i, $\sigma(xy)$, $\sigma(xz)$, and $\sigma(yz)$ operations.

H H
C=C
H H

b.

E: $\begin{vmatrix} 1 & 0 & 0 \\ 0 & 1 & 0 \\ 0 & 0 & 1 \end{vmatrix}$ $C_2(z)$: $\begin{vmatrix} -1 & 0 & 0 \\ 0 & -1 & 0 \\ 0 & 0 & 1 \end{vmatrix}$ $C_2(y)$: $\begin{vmatrix} -1 & 0 & 0 \\ 0 & 1 & 0 \\ 0 & 0 & -1 \end{vmatrix}$

$C_2(x)$: $\begin{vmatrix} 1 & 0 & 0 \\ 0 & -1 & 0 \\ 0 & 0 & -1 \end{vmatrix}$ i: $\begin{vmatrix} -1 & 0 & 0 \\ 0 & -1 & 0 \\ 0 & 0 & -1 \end{vmatrix}$ $\sigma(xy)$: $\begin{vmatrix} 1 & 0 & 0 \\ 0 & 1 & 0 \\ 0 & 0 & -1 \end{vmatrix}$

$\sigma(xz)$: $\begin{vmatrix} 1 & 0 & 0 \\ 0 & -1 & 0 \\ 0 & 0 & 1 \end{vmatrix}$ $\sigma(yz)$: $\begin{vmatrix} -1 & 0 & 0 \\ 0 & 1 & 0 \\ 0 & 0 & 1 \end{vmatrix}$

c.

	E	$C_2(z)$	$C_2(y)$	$C_2(x)$	i	$\sigma(xy)$	$\sigma(xz)$	$\sigma(yz)$
χ	3	–1	–1	–1	–3	1	1	1

d.

Γ_1	1	–1	–1	1	–1	1	1	–1	matching B_{3u}
Γ_2	1	–1	1	–1	–1	1	–1	1	matching B_{2u}
Γ_3	1	1	–1	–1	–1	–1	1	1	matching B_{1u}

e. $\Gamma_1 \times \Gamma_2 = 1 \times 1 + (-1) \times (-1) + (-1) \times 1 + 1 \times (-1) + (-1) \times (-1) + 1 \times 1$
$+ 1 \times (-1) + (-1) \times 1 = 0$

$\Gamma_1 \times \Gamma_3 = 1 \times 1 + (-1) \times 1 + (-1) \times (-1) + 1 \times (-1) + (-1) \times (-1) + 1 \times (-1)$
$+ 1 \times 1 + (-1) \times 1 = 0$

$\Gamma_2 \times \Gamma_3 = 1 \times 1 + (-1) \times 1 + 1 \times (-1) + (-1) \times (-1) + (-1) \times (-1) + 1 \times (-1)$
$+ (-1) \times 1 + 1 \times 1 = 0$

4.20 **a.** $h = 8$ (the total number of symmetry operations)

b. $A_1 \times E = 1 \times 2 + 2 \times 1 \times 0 + 1 \times (-2) + 2 \times 1 \times 0 + 2 \times 1 \times 0 = 0$
$A_2 \times E = 1 \times 2 + 2 \times 1 \times 0 + 1 \times (-2) + 2 \times (-1) \times 0 + 2 \times (-1) \times 0 = 0$
$B_1 \times E = 1 \times 2 + 2 \times (-1) \times 0 + 1 \times (-2) + 2 \times 1 \times 0 + 2 \times (-1) \times 0 = 0$
$B_2 \times E = 1 \times 2 + 2 \times (-1) \times 0 + 1 \times (-2) + 2 \times (-1) \times 0 + 2 \times 1 \times 0 = 0$

c. E: $4 + 2 \times 0 + 4 + 2 \times 0 + 2 \times 0 = 8$
A_1: $1 + 2 \times 1 + 1 + 2 \times 1 + 2 \times 1 = 8$
A_2: $1 + 2 \times 1 + 1 + 2 \times 1 + 2 \times 1 = 8$
B_1: $1 + 2 \times 1 + 1 + 2 \times 1 + 2 \times 1 = 8$
B_2: $1 + 2 \times 1 + 1 + 2 \times 1 + 2 \times 1 = 8$

d. $\Gamma_1 = 2A_1 + B_1 + B_2 + E$:
A_1: $1/8[1 \times 6 + 2 \times 1 \times 0 + 1 \times 2 + 2 \times 1 \times 2 + 2 \times 1 \times 2] = 2$
A_2: $1/8[1 \times 6 + 2 \times 1 \times 0 + 1 \times 2 + 2 \times (-1) \times 2 + 2 \times (-1) \times 2] = 0$
B_1: $1/8[1 \times 6 + 2 \times (-1) \times 0 + 1 \times 2 + 2 \times 1 \times 2 + 2 \times (-1) \times 2] = 1$
B_2: $1/8[1 \times 6 + 2 \times (-1) \times 0 + 1 \times 2 + 2 \times (-1) \times 2 + 2 \times 1 \times 2] = 1$
E: $1/8[2 \times 6 + 2 \times 0 \times 0 + (-2) \times 2 + 2 \times 0 \times 2 + 2 \times 0 \times 2] = 1$

$\Gamma_2 = 3A_1 + 2A_2 + B_1$:
A_1: $1/8[1 \times 6 + 2 \times 1 \times 4 + 1 \times 6 + 2 \times 1 \times 2 + 2 \times 1 \times 0] = 3$
A_2: $1/8[1 \times 6 + 2 \times 1 \times 4 + 1 \times 6 + 2 \times (-1) \times 2 + 2 \times (-1) \times 0] = 2$
B_1: $1/8[1 \times 6 + 2 \times (-1) \times 4 + 1 \times 6 + 2 \times 1 \times 2 + 2 \times (-1) \times 0] = 1$
B_2: $1/8[1 \times 6 + 2 \times (-1) \times 4 + 1 \times 6 + 2 \times (-1) \times 2 + 2 \times 1 \times 0] = 0$
E: $1/8[2 \times 6 + 2 \times 0 \times 4 + (-2) \times 6 + 2 \times 0 \times 2 + 2 \times 0 \times 0] = 0$

4.21 C_{3v}
$\Gamma_1 = 3A_1 + A_2 + E$:
A_1: $1/6[1 \times 6 + 2 \times 1 \times 3 + 3 \times 1 \times 2] = 3$
A_2: $1/6[1 \times 6 + 2 \times 1 \times 3 + 3 \times (-1) \times 2] = 1$
E: $1/6[2 \times 6 + 2 \times (-1) \times 3 + 3 \times 0 \times 2] = 1$

$\Gamma_2 = A_2 + E$:
A_I: $1/6[1 \times 5 + 2 \times 1 \times (-1) + 3 \times 1 \times (-1)] = 0$
A_2: $1/6[1 \times 5 + 2 \times 1 \times (-1) + 3 \times (-1) \times (-1)] = 1$
E: $1/6[2 \times 5 + 2 \times (-1) \times (-1) + 3 \times 0 \times (-1)] = 2$

O_h

$\Gamma_3 = A_{1g} + E_g + T_{1u}$:

A_{1g}: 1/48[6 + 0 + 0 + 12 + 6 + 0 + 0 + 0 + 12 + 12] = 1

A_{2g}: 1/48[6 + 0 + 0 – 12 + 6 + 0 + 0 + 0 + 12 – 12] = 0

E_g: 1/48[12 + 0 + 0 + 0 + 12 + 0 + 0 + 0 + 24 + 0] = 1

T_{1g}: 1/48[18 + 0 + 0 + 12 – 6 + 0 + 0 + 0 – 12 – 12] = 0

T_{2g}: 1/48[18 + 0 + 0 – 12 – 6 + 0 + 0 + 0 – 12 + 12] = 0

A_{1u}: 1/48[6 + 0 + 0 + 12 + 6 + 0 + 0 + 0 – 12 – 12] = 0

A_{2u}: 1/48[6 + 0 + 0 – 12 + 6 + 0 + 0 + 0 – 12 + 12] = 0

E_u: 1/48[12 + 0 + 0 + 0 + 12 + 0 + 0 + 0 – 24 + 0] = 0

T_{1u}: 1/48[18 + 0 + 0 + 12 – 6 + 0 + 0 + 0 + 12 + 12] = 1

T_{2u}: 1/48[18 + 0 + 0 – 12 – 6 + 0 + 0 + 0 + 12 – 12] = 0

4.22

The d_{xy} characters match the characters in the B_{2g} representation.

The $d_{x^2-y^2}$ characters match those of the B_{1g} representation.

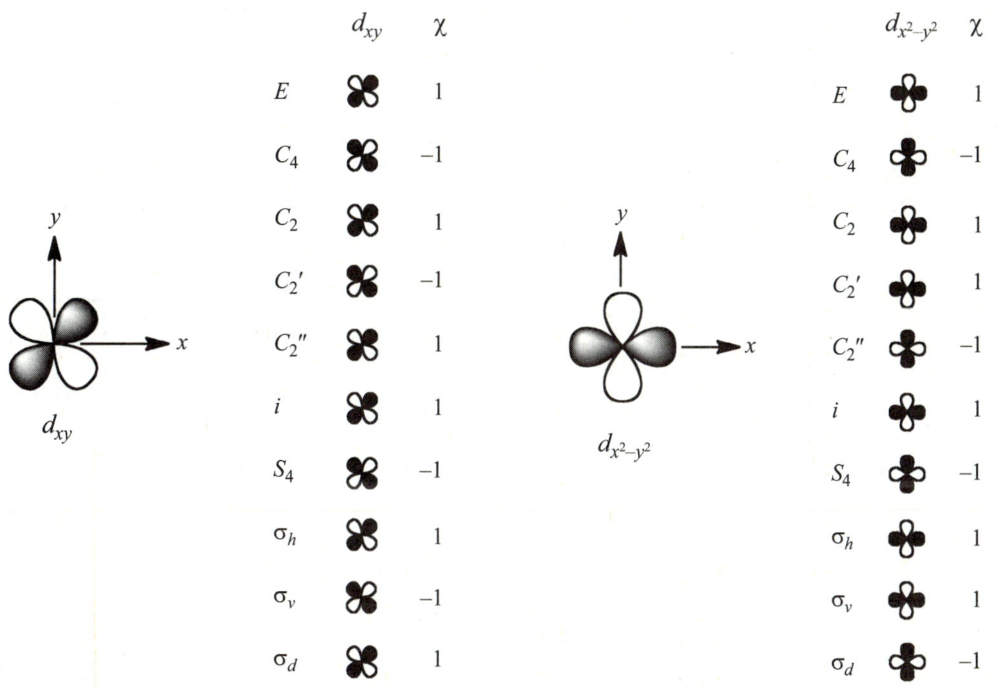

4.23 Chiral: **4.5**: O_2F_2, $[Cr(C_2O_4)_3]^{3-}$ **4.6**: none **4.7**: screw, Captain Hook **4.8**: recycle symbol **4.9**: set of three wind turbine blades, coiled spring

4.24 **a.** Point group: C_{4v}

C_{4v}	E	$2C_4$	C_2	$2\sigma_v$	$2\sigma_d$	
Γ	18	2	–2	4	2	
A_1	1	1	1	1	1	z
A_2	1	1	1	–1	–1	R_z
B_1	1	–1	1	1	–1	
B_2	1	–1	1	–1	1	
E	2	0	–2	0	0	$(x, y), (R_x, R_y)$

b. $\Gamma = 4\,A_1 + A_2 + 2B_1 + B_2 + 5E$

c. Translation: $A_1 + E$ (match x, y, and z)
Rotation: $A_2 + E$ (match R_x, R_y, and R_z)
Vibration: all that remain: $3\,A_1 + 2B_1 + B_2 + 3E$

d. The character for each symmetry operation for the Xe–O stretch is +1. This corresponds to the A_1 irreducible representation, which matches the function z and is therefore IR-active.

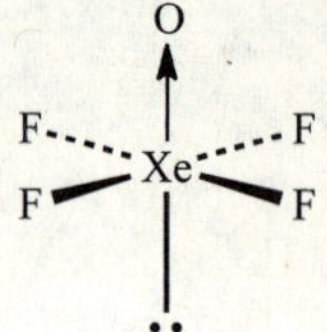

4.25 For SF_6, the axes of the sulfur should point at three of the fluorines. The fluorine axes can be chosen in any way, as long as one from each atom is directed toward the sulfur atom. There are seven atoms with three axes each, for a total of 21.

O_h	E	$8C_3$	$6C_2$	$6C_4$	$3C_2$	i	$6S_4$	$8S_6$	$3\sigma_h$	$6\sigma_d$		
Γ	21	0	–1	3	–3	–3	–1	0	5	3		
T_{1u}	3	0	–1	1	–1	–3	–1	0	1	1	(x,y,z)	
T_{1g}	3	0	–1	1	–1	3	1	0	–1	–1	(R_x, R_y, R_z)	
A_{1g}	1	1	1	1	1	1	1	1	1	1		
A_{2u}	1	1	–1	–1	1	1	–1	1	1	–1		
E_g	2	–1	0	0	2	2	0	–1	2	0		$(2z^2 - x^2 - y^2, x^2 - y^2)$
T_{2u}	3	0	1	–1	–1	–3	1	0	1	–1		
T_{2g}	3	0	1	–1	–1	3	–1	0	–1	1		(xy, xz, yz)

Reduction of Γ gives $\Gamma = 3T_{1u} + T_{1g} + A_{1g} + E_g + T_{2g} + T_{2u}$. T_{1u} accounts for translation and also infrared active vibrational modes. T_{1g} is rotation. The remainder are vibrations that are infrared inactive.

4.26 **a.** *cis*-$Fe(CO)_4Cl_2$ has C_{2v} symmetry.

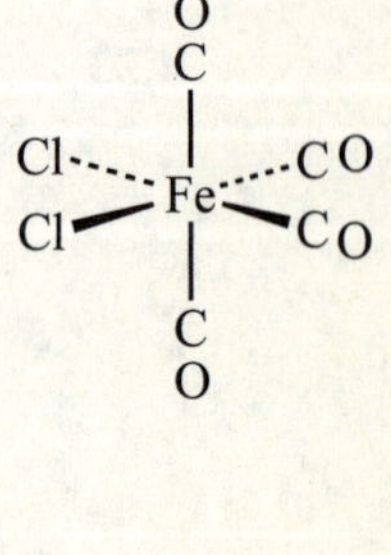

The vectors for CO stretching have the representation Γ below:

C_{2v}	E	C_2	$\sigma_v(xz)$	$\sigma_v'(yz)$	
Γ	4	0	2	2	
A_1	1	1	1	1	z
A_2	1	1	–1	–1	
B_1	1	–1	1	–1	x
B_2	1	–1	–1	1	y

$n(A_1) = 1/4[4 \times 1 + 0 \times 1 + 2 \times 1 + 2 \times 1] = 2$
$n(A_2) = 1/4[4 \times 1 + 0 \times 1 + 2 \times (-1) + 2 \times (-1)] = 0$
$n(B_1) = 1/4[4 \times 1 + 0 \times (-1) + 2 \times 1 + 2 \times (-1)] = 1$
$n(B_2) = 1/4[4 \times 1 + 0 \times (-1) + 2 \times (-1) + 2 \times 1] = 1$
$\Gamma = 2\,A_1 + B_1 + B_2$, all four IR active.

b. *trans*-$Fe(CO)_4Cl_2$ has D_{4h} symmetry.

D_{4h}	E	$2C_4$	C_2	$2C_2'$	$2C_2''$	i	$2S_4$	σ_h	$2\sigma_v$	$2\sigma_d$	
Γ	4	0	0	2	0	0	0	4	2	0	
A_{2u}	1	1	1	–1	–1	–1	–1	–1	1	1	z
E_u	2	0	–2	0	0	–2	0	2	0	0	(x,y)

Omitting the operations that have zeroes in Γ:
$n(A_{2u}) = 1/16[4 \times 1 + 2 \times 2 \times (-1) + 4 \times (-1) + 2 \times 2 \times 1] = 0$
$n(E_u) = 1/16[4 \times 2 + 2 \times 2 \times 0 + 4 \times 2 + 2 \times 2 \times 0] = 1$ (IR active)

Note: In checking for IR-active bands, it is only necessary to check the irreducible representations having the same symmetry as x, y, or z, or a combination of them.

c. $Fe(CO)_5$ has D_{3h} symmetry.

The vectors for C–O stretching have the following representation Γ:

D_{3h}	E	$2C_3$	$3C_2$	σ_h	$2S_3$	$3\sigma_v$	
Γ	5	2	1	3	0	3	
E'	2	–1	0	2	–1	0	(x, y)
A_2''	1	1	–1	–1	–1	1	z

$n(E') = 1/12\ [(5 \times 2) + (2 \times 2 \times -1) + (3 \times 2)] = 1$
$n(A_2'') = 1/12\ [(5 \times 1) + (2 \times 2 \times 1) + (3 \times 1 \times -1) + (3 \times -1) + (3 \times 3 \times 1)] = 1$

There are two bands, one matching E' and one matching A_2''. These are the only irreducible representations that match the coordinates x, y, and z.

4.27 From problem 4.24 we note that the irreducible representations for vibration are $3A_1 + 2B_1 + B_2 + 3E$. From the C_{4v} character table we note that each of these representations matches Raman-active functions: $A_1\,(x^2 + y^2, z^2)$; $B_1\,(x^2 - y^2)$; $B_2\,(xy)$; and $E\,(xz, yz)$, so all are Raman-active.

4.28 **a.** The point group is C_{2h}.

b. Using the Si–I bond vectors as a basis generates the representation:

C_{2h}	E	C_2	i	σ_h		
Γ	4	0	0	0		
A_g	1	1	1	1		x^2+y^2, z^2
B_g	1	−1	1	−1		xz, yz
A_u	1	1	−1	−1	z	
B_u	1	−1	−1	1	x, y	

$$\Gamma = A_g + B_g + A_u + B_u$$

The A_u and B_u vibrations are infrared active.

c. The A_g and B_g vibrations are Raman active.

4.29 *trans* isomer (D_{4h}): *cis* isomer (C_{2v}):

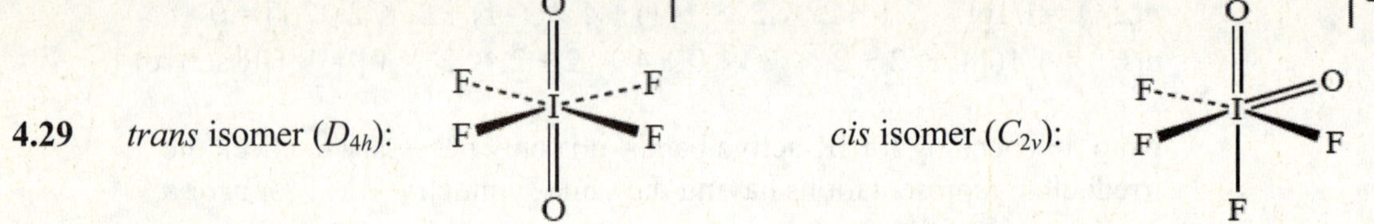

trans:

D_{4h}	E	$2C_4$	C_2	$2C_2'$	$2C_2''$	i	$2S_4$	σ_h	$2\sigma_v$	$2\sigma_d$	
Γ	2	2	2	0	0	0	0	0	2	2	
A_{1g}	1	1	1	1	1	1	1	1	1	1	
A_{2u}	1	1	1	−1	−1	−1	−1	−1	1	1	z

There is only a single IR-active I–O stretch (the antisymmetric stretch), A_{2u}.

cis:

C_{2v}	E	C_2	$\sigma(xz)$	$\sigma(yz)$	
Γ	2	0	2	0	
A_1	1	1	1	1	z
B_1	1	−1	1	−1	x

There are two IR-active I–O stretches, the A_1 and B_1 (symmetric and antisymmetric). Infrared spectra should therefore be able to distinguish between these isomers. (Reversing the x and y axes would give $A_1 + B_2$. Because B_2 matches y, it would also represent an IR-active vibration.)

Because these isomers would give different numbers of IR-active absorptions, infrared spectra should be able to distinguish between them—as it did in the reference.

4.30 **I** has C_2 symmetry, with a C_2 axis running right to left, perpendicular to the Cl, N, Cl, N and Cl, P, Cl, P faces.
II also has C_2 symmetry, with the same C_2 axis as **I**.
III has only an inversion center and C_i symmetry.

4.31 An example for each of the five possible point groups:

T_d: 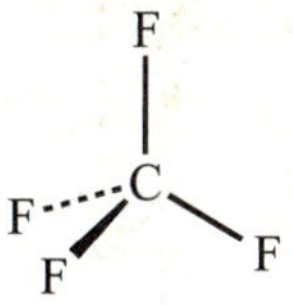 C_{3v}: 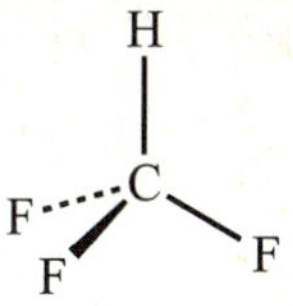 C_{2v}: 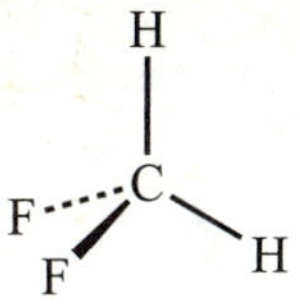 C_s: 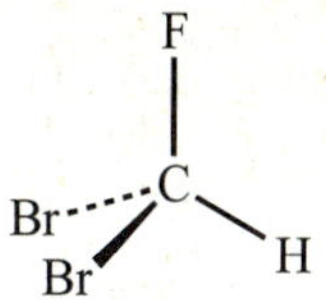 C_1:

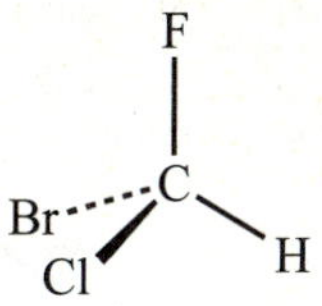

4.32 **a.** The S–C–C portion is linear, so the molecule has a C_3 axis along the line of these three atoms, three σ_v planes through these atoms and an F atom on each end, but no other symmetry elements. $\boldsymbol{C_{3v}}$.

b. The molecule has only an inversion center, so it is $\boldsymbol{C_i}$. The inversion center is equivalent to an S_2 axis perpendicular to the average plane of the ring.

c. $M_2Cl_6Br_4$ also is $\boldsymbol{C_i}$.

d. This complex has a C_3 axis, splitting the three N atoms and the three P atoms (almost as drawn), but no other symmetry elements. $\boldsymbol{C_3}$.

e. The most likely isomer has the less electronegative Cl atoms in axial positions. Point group: $\boldsymbol{C_{2v}}$.

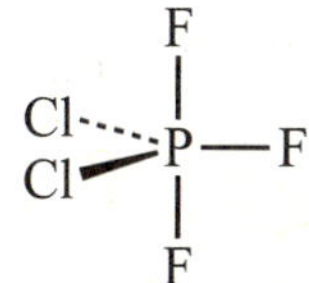

4.33 **a.** C_{3v}

b. D_{5h}

4.34 Web of Science and SciFinder Scholar should be helpful, but simply searching for these symmetries using a general search should provide examples of these point groups. Some examples:

a. S_6: $Mo_2(SC_6H_2Me_3)_6$ (M. H. Chisholm, J. F. Corning, and J. C. Huffman, *J. Am. Chem. Soc.*, **1983**, *105*, 5924)
$Mo_2(NMe_2)_6$ (M. H. Chisholm, R. A. Cotton, B. A. Grenz, W. W. Reichert, L. W. Shive, and B. R. Stults, *J. Am. Chem. Soc.*, **1976**, *98*, 4469)
$[NaFe_6(OMe)_{12}(dbm)_6]^+$ (dbm = dibenzoylmethane, $C_6H_5COCCOC_6H_5$) (F. L. Abbati, A. Cornia, A. C. Fabretti, A. Caneschi, and D. Garreschi, *Inorg. Chem.*, **1998**, *37*, 1430)

b. T $Pt(CF_3)_4$, C_{44}

c. I_h C_{20}, C_{80}

d. T_h $[Co(NO_2)_6]^{3-}$, $Mo(NMe_2)_6$

In addition to examples that can be found using Web of Science, SciFinder, and other Internet search tools, numerous examples of these and other point groups can be found in I. Hargittai and M. Hargittai, *Symmetry Through the Eyes of a Chemist,* as listed in the General References section.

CHAPTER 5: MOLECULAR ORBITALS

5.1 There are three possible bonding interactions

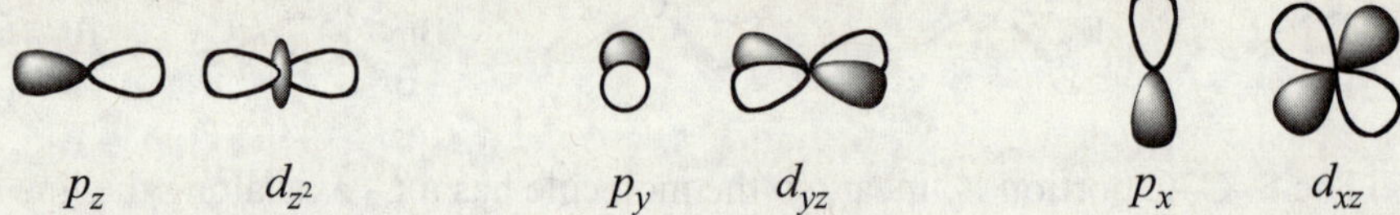

5.2 **a.** Li_2 has a bond order of 1.0 (two electrons in a σ bonding orbital; see Figures 5.7 and 5.1). Li_2^+ has a bond order of only 0.5 (one electron in a σ bonding orbital). Therefore, Li_2 has the shorter bond.

b. F_2 has a bond order of 1.0 (see Figure 5.7). F_2^+ has one less antibonding (π*) electron and as a result has a higher bond order, 1.5. F_2^+ would therefore be expected to have the shorter bond.

c. Expected bond orders (see Figure 5.1):

	Bonding electrons	Antibonding electrons	Bond order
He_2^+	2	1	$\frac{1}{2}(2-1) = 0.5$
HHe^+	1	1	$\frac{1}{2}(1-1) = 0$
H_2^+	1	0	$\frac{1}{2}(2-1) = 0.5$

Both He_2^+ and H_2^+ have bond orders of 0.5. H_2^+ would be expected to have the shorter bond because hydrogen atoms are smaller than helium atoms.

5.3 **a.** These diatomic molecules should have similar bond orders to the analogous diatomics from the row directly above them in the periodic table:

P_2	bond order = 3 (like N_2)	
S_2	bond order = 2 (like O_2)	
Cl_2	bond order = 1 (like F_2)	Cl_2 has the weakest bond.

b. The bond orders match those of the analogous oxygen species (page 137):

S_2^+	bond order = 2.5	
S_2	bond order = 2	
S_2^-	bond order = 1.5	S_2^- has the weakest bond.

c. Bond orders:

NO^+	bond order = 3 (isoelectronic with CO, Figure 5.14)
NO	bond order = 2.5 (one more (antibonding) electron than CO)
NO^-	bond order = 2 (two more (antibonding) electrons than CO)

NO^- has the lowest bond order and therefore the weakest bond.

5.4

O_2^{2-} has a single bond, with four electrons in the π^* orbitals canceling those in the π orbitals.

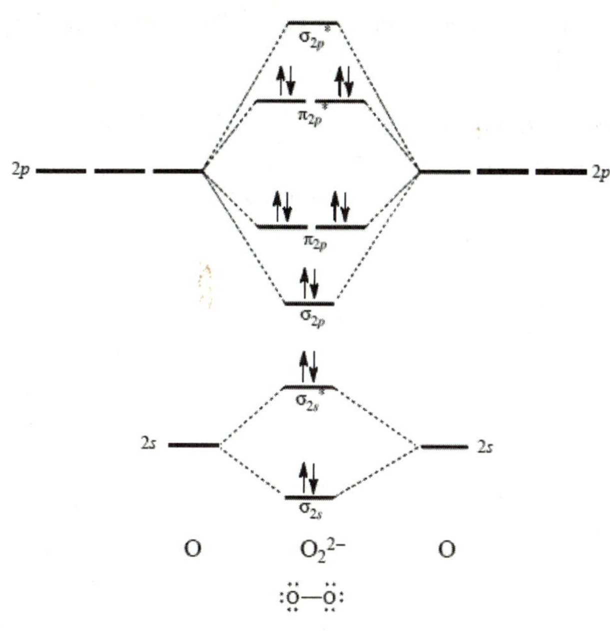

O_2^- has three electrons in the π^* orbitals, and a bond order of 1.5. The Lewis structures have an unpaired electron and an average bond order of 1.5.

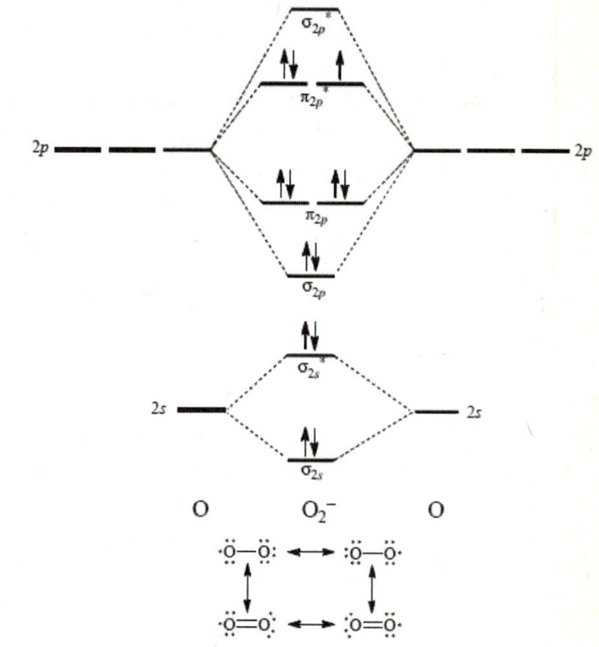

O_2 has two unpaired electrons in its π^* orbitals, and a bond order of 2. The simple Lewis structure has all electrons paired, which does not match the paramagnetism observed experimentally.

Bond lengths are therefore in the order $O_2^{2-} > O_2^- > O_2$, and bond strengths are the reverse of this order.

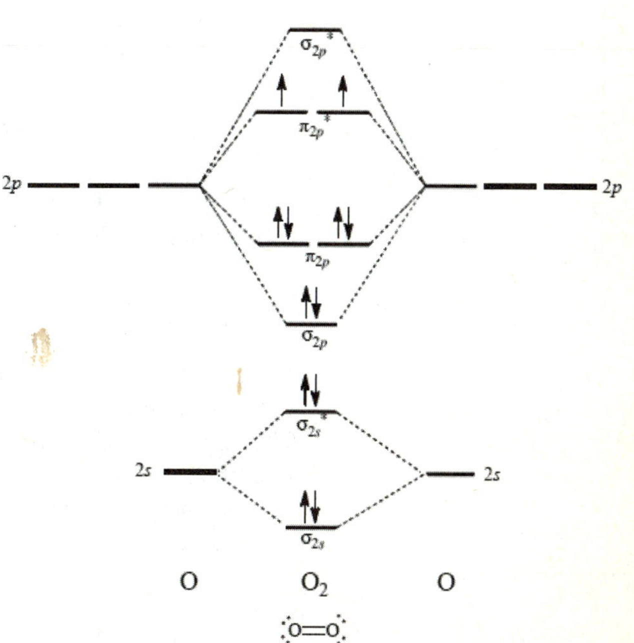

5.5

	Bond Order (Figures 5.5 and 5.7)	Bond Distance (pm)	Unpaired Electrons
C_2^{2-}	3	119	0
N_2^{2-}	2	122.4	2
O_2^{2-}	1	149 (very long)	0
O_2	2	120.7	2

The bond distance in N_2^{2-} is very close to the expected bond distance for a diatomic with 12 valence electrons, as shown in Figure 5.8.

5.6 The energy level pattern would be similar to the one shown in Figure 5.5, with the interacting orbitals the 3*s* and 3*p* rather than 2*s* and 2*p*. All molecular orbitals except the highest would be occupied by electron pairs, and the highest orbital (σ_u*) would be singly occupied, giving a bond order of 0.5. Because the bond in Ar_2^+ would be weaker than in Cl_2, the Ar–Ar distance would be expected to be longer (calculated to be > 300 pm; see the reference).

5.7 **a.** The energy level diagram for NO is on the right. The odd electron is in a π_{2p}* orbital

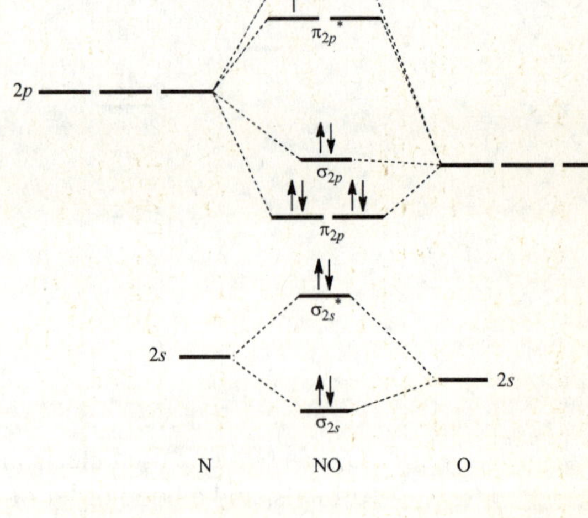

b. O is more electronegative than N, so its orbitals are slightly lower in energy. The bonding orbitals are slightly more concentrated on O.

c. The bond order is 2.5, with one unpaired electron

d.

NO^+	Bond order = 3	shortest bond (106 pm)
NO	Bond order = 2.5	intermediate (115 pm)
NO^-	Bond order = 2	longest bond (127 pm), two electrons in antibonding orbitals.

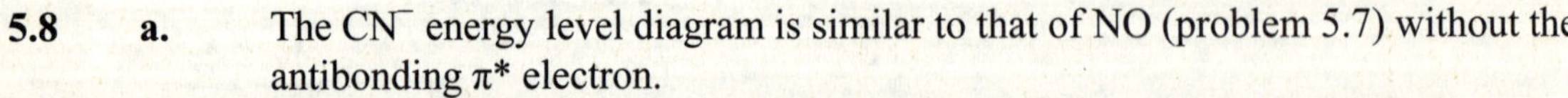

5.8 **a.** The CN^- energy level diagram is similar to that of NO (problem 5.7) without the antibonding π* electron.

b. The bond order is three, with no unpaired electrons.

c. The HOMO is the σ_{2p} orbital, which can interact with the 1*s* of the H^+, as in the diagram at right. The bonding orbital has an energy near that of the π orbitals; the antibonding orbital becomes the highest energy orbital.

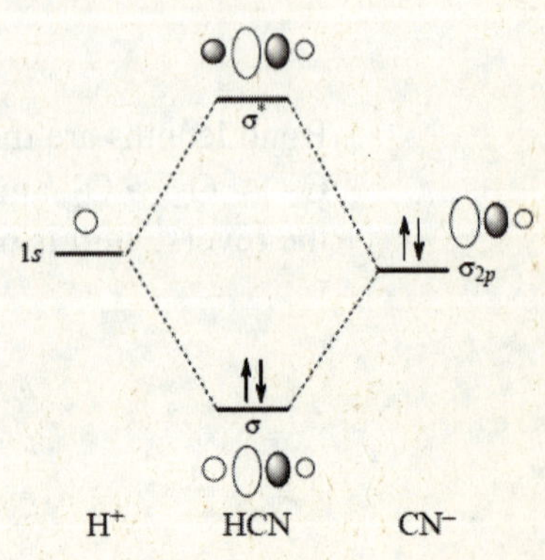

5.9 **a.** OF^- has 14 valence electrons, four in the π_{2p}* orbitals (see the diagram in the answer to problem 5.7).

b. The net result is a single bond between two very electronegative atoms, and no unpaired electrons.

c. The concentration of electrons in the π* orbital is more on the O, so combination with the positive proton at that end is more likely. In fact, H^+ bonds to the oxygen atom, at an angle of 97°, as if the bonding were through a p orbital on O.

5.10 The molecular orbital description of KrF^+ would predict that this ion, which has the same number of valence electrons as F_2, would have a single bond. KrF_2 would also be expected, on the basis of the VSEPR approach, to have single Kr–F bonds, in addition to three lone pairs on Kr. Reported Kr–F distances: KrF^+: 176.5-178.3 pm; KrF_2: 186.8-188.9 pm. The presence of lone pairs in KrF_2 may account for the longer bond distances in this compound.

5.11 **a.** The $KrBr^+$ energy level diagram is at the right.

b. The HOMO is polarized toward Br, since its energy is closer to that of the Br 4p orbital.

c. Bond order = 1

d. Kr is more electronegative. Its greater nuclear charge exerts a stronger pull on the shared electrons.

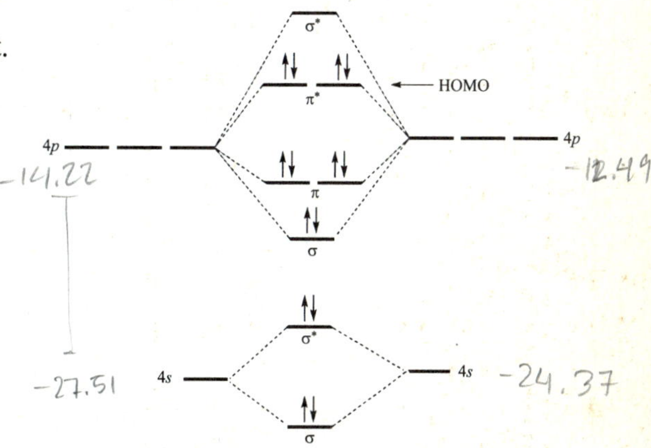

5.12 The molecular orbitals for SH^- are given below. Net, one bond.

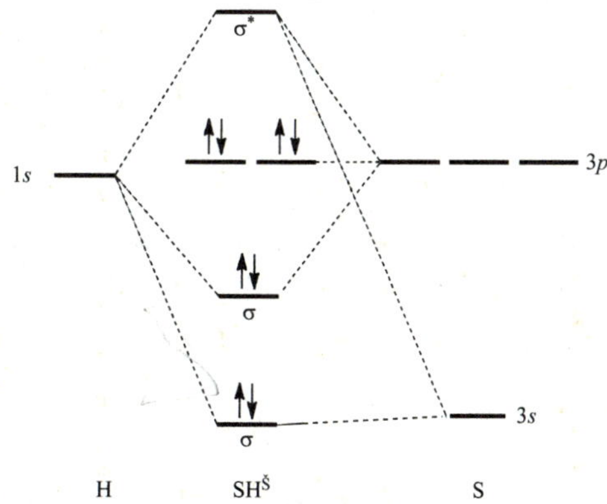

S orbital energies are –22.7 eV (3s) and –11.6 eV (3p); the 1s of H has an energy of –13.6 eV. Because of the difference in their atomic orbital energies, the 1s orbital of hydrogen and the 3s orbital of sulfur interact only weakly; this is shown in the diagram by a slight stabilization of the lowest energy molecular orbital with respect to the 3s orbital of sulfur.

5.13 **a.** The group orbitals on the hydrogen atoms are

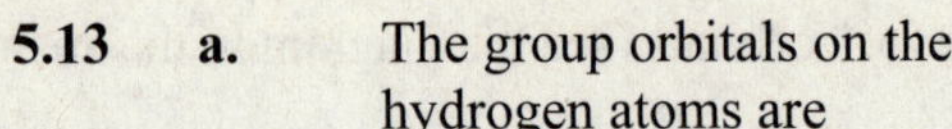

and

The first group orbital interacts with the 2*s* orbital on carbon:

And the second group orbital interacts with a 2*p* orbital on carbon:

Carbon's remaining 2*p* orbitals are nonbonding.

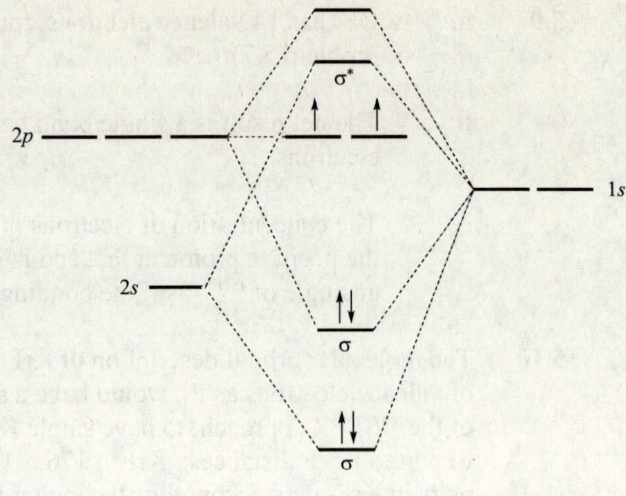

b. CH_2 is a paramagnetic diradical, with one electron in each of the p_x and p_y orbitals of carbon. (A bent singlet state, with all electrons paired, is also known, with a calculated bond angle of approximately 130°.)

5.14 **a.** BeH_2

Group Orbitals: H Be H H Be H

Be Orbitals with Matching Symmetry: 2*s* $2p_z$

MO Diagram:

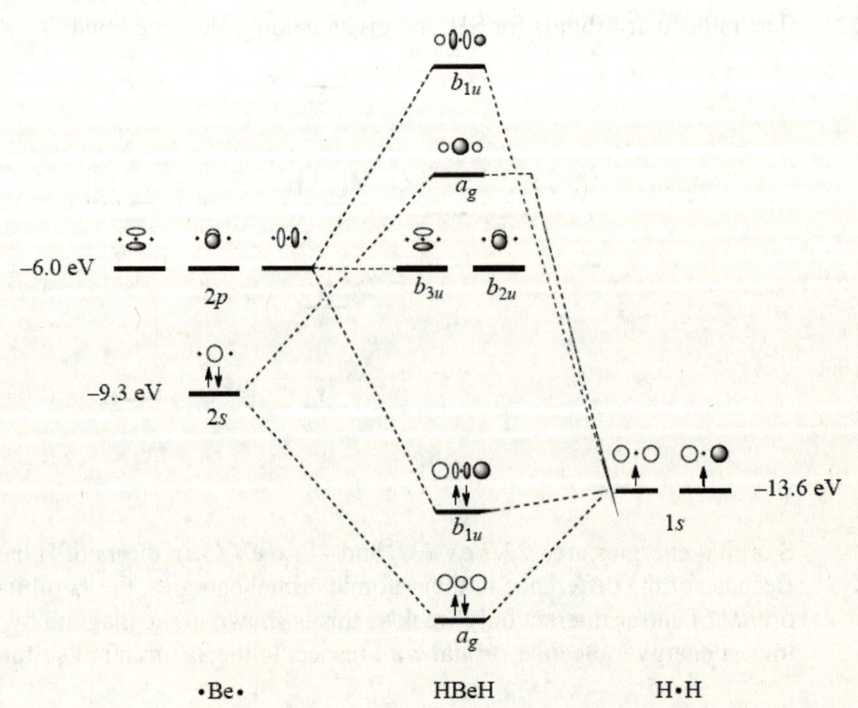

5.15 BeF_2 uses *s* and *p* orbitals on all three atoms, and is isoelectronic with CO_2. The energy level diagram for CO_2 in Figure 5.26 can be used as a guide, with the orbitals of Be higher in energy than those of C and the orbitals of F lower in energy than those of O. Calculated molecular orbital shapes are shown below. The CO_2 orbitals in Figure 5.26 can be compared with these.

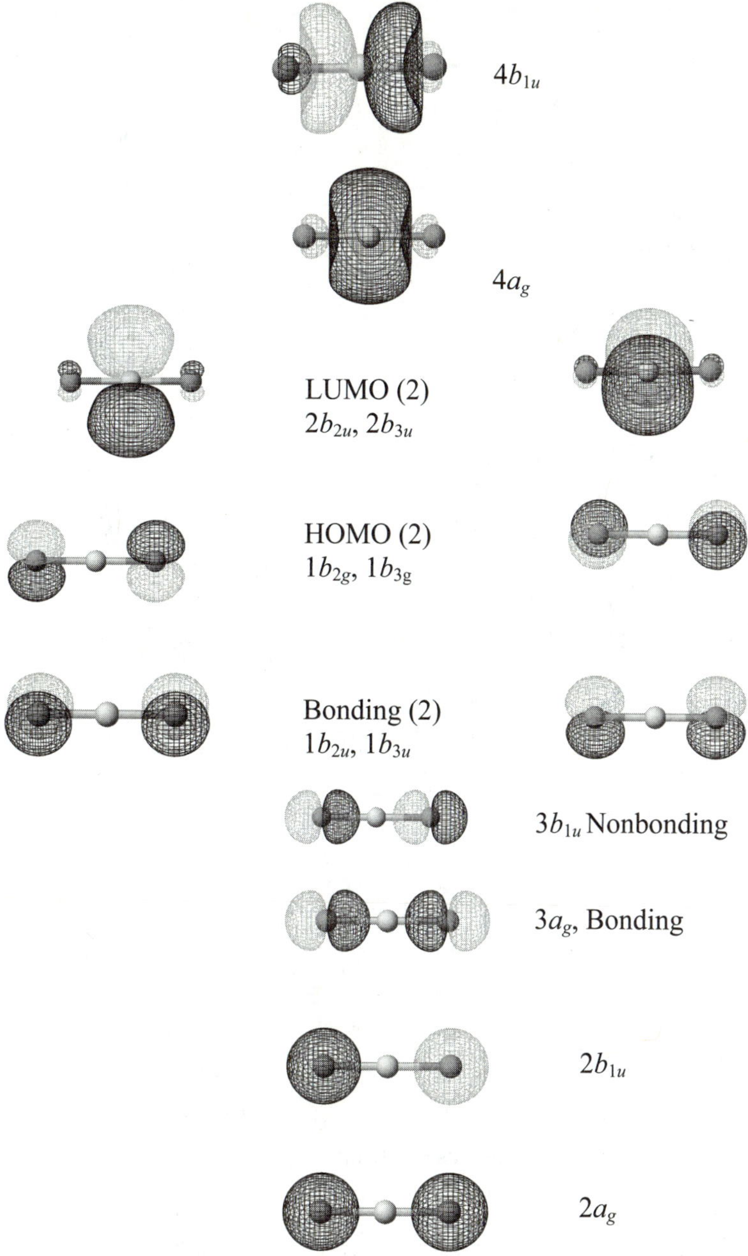

Of the occupied orbitals, there are three bonding (two π and one σ) and five nonbonding (two π and three σ). (References: W. R. Wadt and W. A. Goddard III, *J. Am. Chem. Soc.*, **1974**, *96*, 5996; R. Gleiter and R. Hoffmann, *J. Am. Chem. Soc.*, **1968**, *90*, 5457; C. W. Bauschlicher, Jr. and I. Shavitt, *J. Am. Chem. Soc.*, **1978**, *100*, 739.)

5.16 **a.** The group orbitals of the fluorines are:

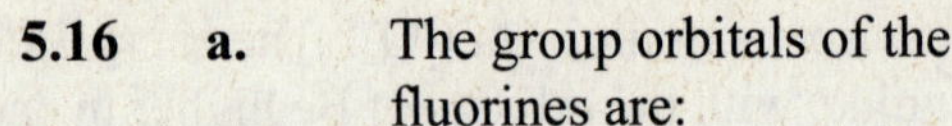

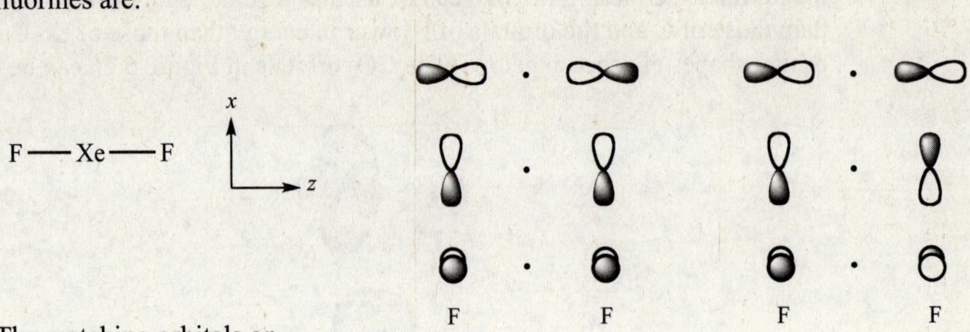

b. The matching orbitals on xenon are:

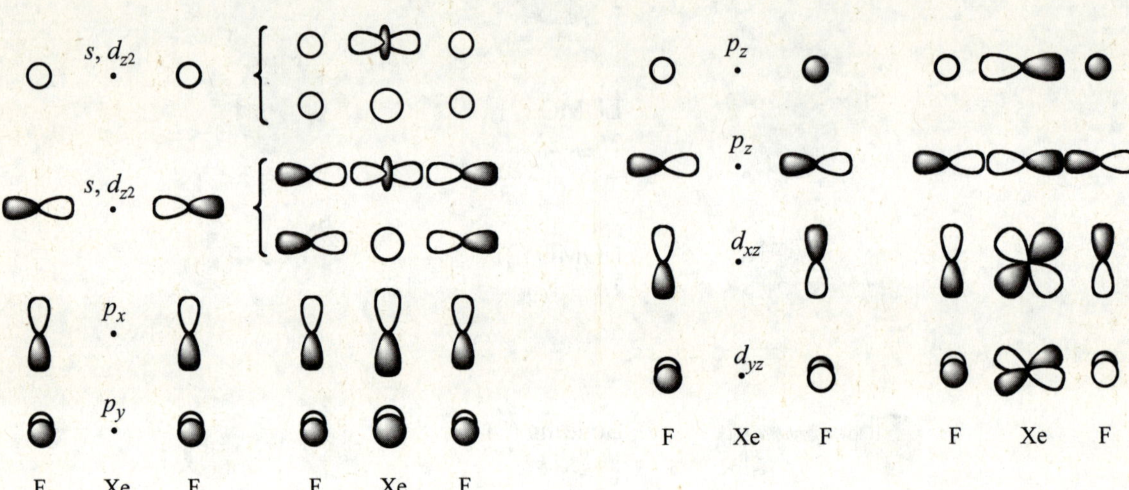

5.17 1. The point group of TaH_5 is C_{4v}.

2. Axes can be assigned as shown:

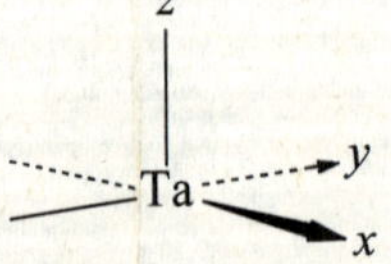

3. Construction of reducible representation:

C_{4v}	E	$2C_4$	C_2	$2\sigma_v$	$2\sigma_d$		
Γ	5	1	1	3	1		
A_1	1	1	1	1	1	z	z^2
A_2	1	1	1	−1	−1	R_z	
B_1	1	−1	1	1	−1		x^2-y^2
B_2	1	−1	1	−1	1		xy
E	2	0	−2	0	0	$(x, y), (R_x, R_y)$	(xz, yz)

4. Γ reduces to $2\,A_1 + B_1 + E$

5, 6. Two group orbitals, shown at right, have A_1 symmetry. These may interact with the p_z, d_{z^2}, and *s* orbitals of Ta.

One group orbital has B_1 symmetry. It Can interact with the $d_{x^2-y^2}$ orbital of Ta.

A degenerate pair of group orbitals has E symmetry. It may interact with the (p_x, p_y) and (d_{xz}, d_{yz}) pairs of Ta.

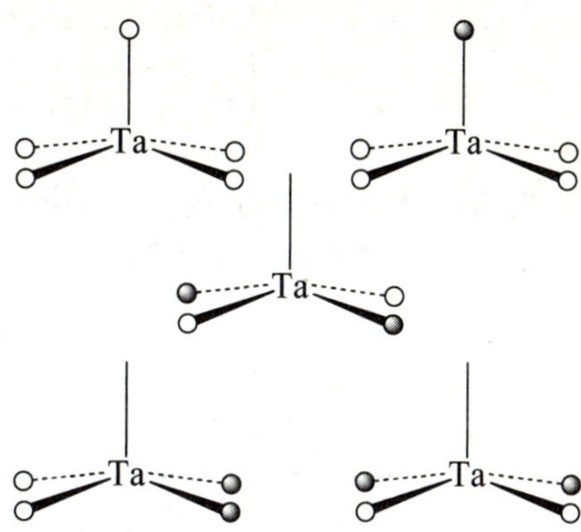

5.18 **a.** The energy level diagram for O_3 with the simple combinations of *s* and *p* orbitals is shown below:

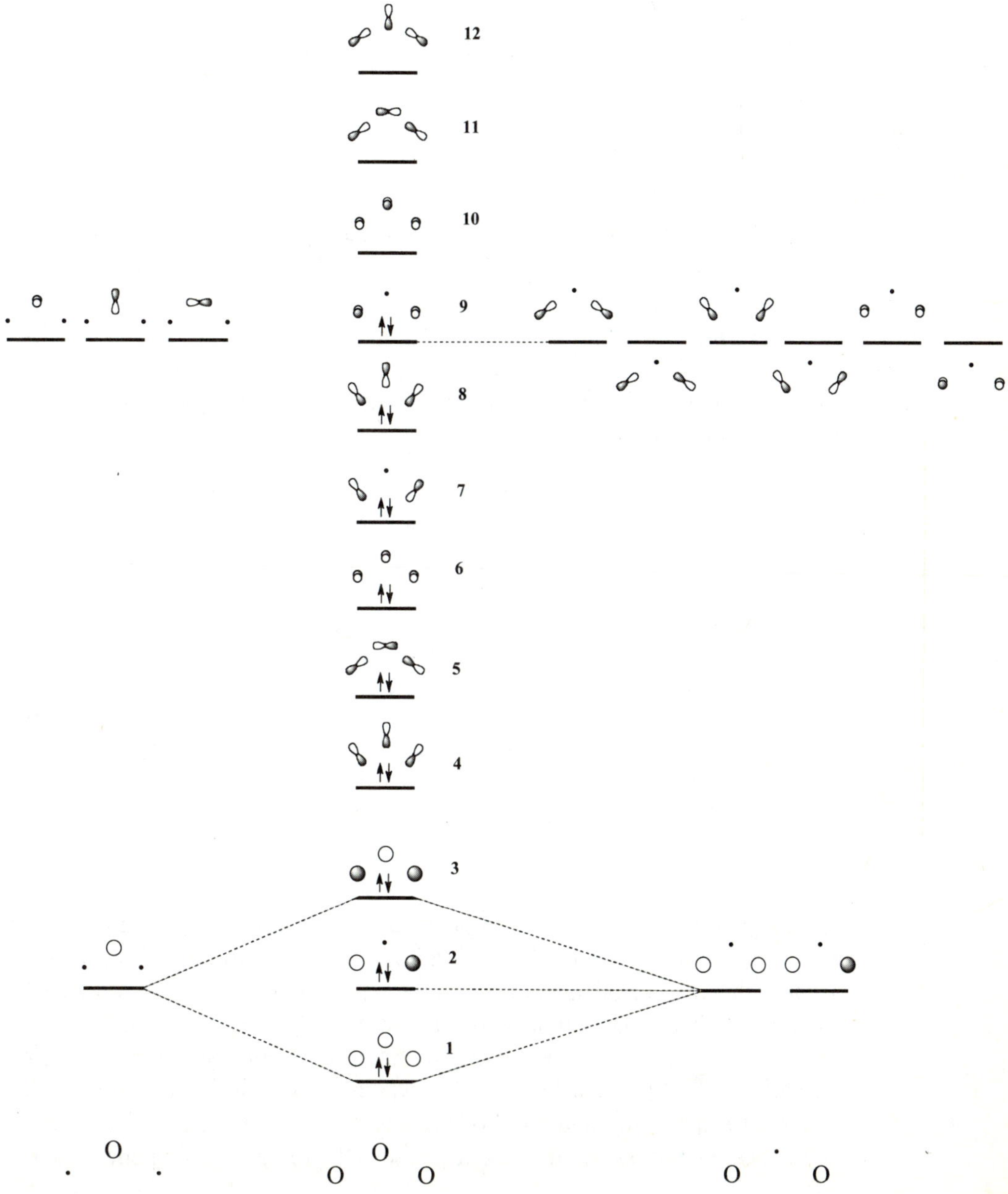

12

11

10

9

8

7

6

5

4

3

2

1

O

O O

O O

b. Mixing of *s* and *p* orbitals is fairly small, showing mostly in the four lowest orbitals. The order of orbitals may vary depending on the calculation method (for example, PM3 and AM1 methods reverse the orders of HOMO and HOMO –1).

5.19 SO_3 has molecular orbitals similar to those of BF_3, described on pp. 165-167. The irreducible representations below are labeled for the oxygen orbitals.

D_{3h}	E	$2C_3$	$3C_2$	σ_h	$2S_3$	$3\sigma_v$	
$\Gamma(s)$	3	0	1	3	0	1	
$\Gamma(p_y)$	3	0	1	3	0	1	
$\Gamma(p_x)\parallel$	3	0	–1	3	0	–1	
$\Gamma(p_z)\perp$	3	0	–1	–3	0	1	
A_1'	1	1	1	1	1	1	
A_2'	1	1	–1	1	1	–1	
A_2''	1	1	–1	–1	–1	1	z
E'	2	–1	0	2	–1	0	(x, y)
E''	2	–1	0	–2	1	0	

σ	$\Gamma(s) = A_1' + E'$	Sulfur s, p_x, and p_y
σ	$\Gamma(p_y) = A_1' + E'$	Sulfur s, p_x, and p_y
$\pi_{\parallel}$	$\Gamma(p_x) = A_2' + E'$	Sulfur p_x and p_y
$\pi_{\perp}$	$\Gamma(p_z) = A_2'' + E''$	Sulfur p_z

5.20 As a cyclic (triangular) ion, H_3^+ has a pair of electrons in a bonding orbital and two vacant orbitals that are slightly antibonding:

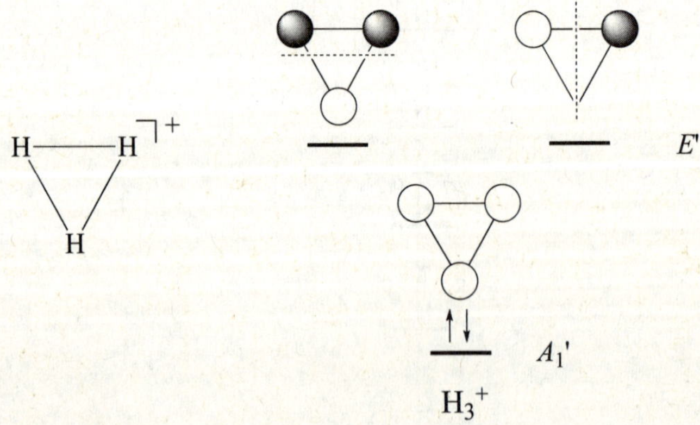

5.21 The thiocyanate ion, SCN^-, has molecular orbitals similar to those of CO_2, but with more mixing between the *s* orbital of C and the *s* and *p* orbitals of S and because the valence orbital potential energies of S are very close to those of C and those of N are only slightly lower. Because the energy match between valence orbitals is good, there is significant double bonding on both sides of the carbon. This is consistent with the resonance structures shown at right, with the top structure favored. For cyanate, OCN^-, the *s* and *p* mix is similar on the N side, but less on the O side because the oxygen orbital energies are much lower in

1– :S=C=N:
1– :S—C≡N:
1+ 2– :S≡C—N:

energy. The structures described in Chapter 3, p. 54, (a mix of two double bonds and O–C and C≡N) fit this ion also. For fulminate, CNO^-, the large difference between the C and O orbital energies makes mixing of the terminal orbitals difficult. As a result of this and the formal charge effects described on pp. 54-55, the bonding in this ion is weak and it is stable only when combined in a complex with a metal ion.

5.22 The highest occupied orbitals in either SCN^- or OCN^- are π non-bonding orbitals (see Figure 5.26 for the similar CO_2 orbitals). Combination with H^+ or with metal ions depends on the energy match of these orbitals with those of the positive ion. The energy of the H orbital matches the energy of the N orbital better than either S or O orbitals. The S and N orbitals are close enough in energy to allow either end to bond with metal ions; slight differences dictate the preferred location. The S can also use the empty 3*d* orbitals to accept π bonding from some metal ions.

5.23 The CN^- molecular orbitals are similar to those of CO (Figure 5.14, p. 146), but with less difference between the C and N atomic orbital energies than between C and O orbitals. As a result, the HOMO should be more evenly balanced between the two atoms, and bonding at both ends seems possible. The Prussian blue structures ($Fe_4[Fe(CN)_6]_3$ or $KFe(Fe(CN)_6)$) have iron and CN^- in arrangements that have both Fe–C and Fe–N bonds.

5.24 **a.** The resonance structures were considered in Problem 3.3, showing bent structures with primarily double bond character in S=N and single bonding in N–O or S–O. SNO^- is more stable on the basis of formal charges.

b. The molecular orbitals should be similar to those of O_3, with more mixing of *s* and *p* orbitals because of the difference between atomic orbital energies of S and O as terminal atoms. The π bonding, nonbonding, and antibonding orbitals are numbers 6, 9, and 10 in the ozone diagram in the answer to Problem 5.18.

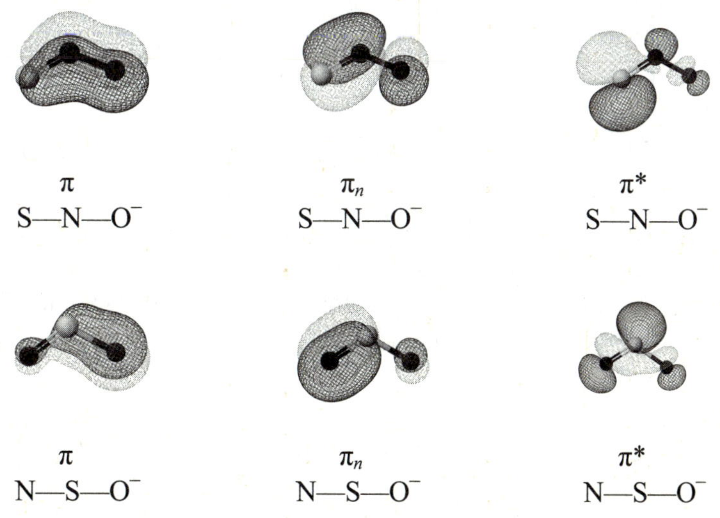

c. The calculated bond distances for these ions are:

ion	N–S	S–O	N–O
SNO^-	171 pm		120 pm
NSO^-	146 pm	149 pm	

NSO^- has the shorter N–S bond and the higher energy N–S stretching vibration.

5.25 SF_4 has C_{2v} symmetry. Treating the four F atoms as simple spherical orbitals, the reducible representation Γ_σ can be found and reduced to $\Gamma_\sigma = 2A_1 + B_1 + B_2$. Overall, the bonding orbitals can be dsp^2 or d^2sp, with the s and p_z or d_{z^2} orbital with A_1 symmetry, p_x or d_{xy} with B_1 symmetry, and p_y or d_{yz} with B_2 symmetry. The p_z or d_{z^2} orbital remaining accounts for the lone pair. (The use of the trigonal bipyramidal hybrids dsp^3 or d^3sp include the lone pair as one of the five locations.)

F
:—S
F
F
F

C_{2v}	E	C_2	$\sigma(xz)$	$\sigma(yz)$	
Γ_σ	4	0	2	2	
A_1	1	1	1	1	z, z^2
A_2	1	1	–1	–1	
B_1	1	–1	1	–1	x, xz
B_2	1	–1	–1	1	y, yz

5.26 A square pyramidal molecule has the reducible representation $\Gamma = E + 2A_1 + B_1$.

C_{4v}	E	$2C_4$	C_2	$2\sigma_v$	$2\sigma_d$	
Γ_σ	5	1	1	3	1	
E	2	0	–2	0	0	$(x, y)\ (xz, yz)$
A_1	1	1	1	1	1	z, z^2, x^2+y^2
B_1	1	–1	1	1	–1	$x^2–y^2$

There appear to be three possibilities for combing orbitals, depending on the details of their relative energies: dsp^3 (p_x and p_y for E, s and p_z for A_1, $d_{x^2-y^2}$ for B_1), d^2sp^2 (substituting d_{z^2} for p_z), and d^3sp (substituting d_{xz} and d_{yz} for p_x and p_y). Although d_{xz} and d_{yz} appear to work, they actually have their electron concentration between the B atoms, and therefore do not participate in σ bonding, so d^3sp or d^2sp^2 fit better.

5.27 Square planar compounds have D_{4h} symmetry.

D_{4h}	E	$2C_4$	C_2	C_2'	C_2''	i	$2S_4$	σ_h	$2\sigma_v$	$2\sigma_d$	
Γ	4	0	0	2	0	0	0	4	2	0	
E_u	2	0	–2	0	0	–2	0	2	0	0	(x,y)
A_{1g}	1	1	1	1	1	1	1	1	1	1	z^2
B_{1g}	1	–1	1	1	–1	1	–1	1	1	–1	$x^2–y^2$

$$\Gamma_\sigma = A_{1g} + B_{1g} + E_u$$
$$\downarrow \quad \downarrow \quad \downarrow$$
$$s, d_{z^2} \; d_{x^2-y^2} \; p_x, p_y$$

dsp^2 hybrids are the usual ones used for square planar compounds, although d^2p^2 is also possible. Since the d_{z^2} orbital does not extend far in the xy plane, it is less likely to participate in σ bonding.

5.28 **a.** PCl_5 has D_{3h} symmetry.

D_{3h}	E	$2C_3$	$3C_2$	σ_h	$2S_3$	$3\sigma_v$	
Γ	5	2	1	3	0	3	
E'	2	–1	0	2	–1	0	(x, y) $(x^2–y^2, xy)$
A_1'	1	1	1	1	1	1	z^2
A_2''	1	1	–1	–1	–1	1	z

$\Gamma = E' + 2A_1' + A_2''$, so the hybrids are dsp^3 or d^3sp.

b. This could also be analyzed separately for the axial and the equatorial positions. The p_z and d_{z^2} orbitals can bond to the axial chlorines ($A_1' + A_2''$) and the s, p_x, and p_y orbitals or the s, $d_{x^2-y^2}$, and d_{xy} orbitals can bond to the equatorial chlorines (E').

c. The d_{z^2} orbital extends farther than the p orbitals, making the axial bonds a bit longer.

5.29 **a.** Cl_2^+ has one fewer electron than Cl_2, so the π* levels have three, rather than four, electrons. As a result, Cl_2^+ has a bond order of 1.5, and the bond is shorter and stronger than that of of Cl_2 (189 pm, compared with 199 pm for Cl_2).

b. Cl_4^+ has such an elongated rectangular shape (194 pm by 294 pm) that it must be essentially a Cl_2 and a Cl_2^+ side by side, with only a weak attraction between them through the π* orbitals. The Cl–Cl bond in Cl_2 is 199 pm long; apparently the weak side-to-side bond draws off some of the antibonding electron density, strengthening and shortening the other two shorter Cl–Cl bonds.

5.30 **a.**

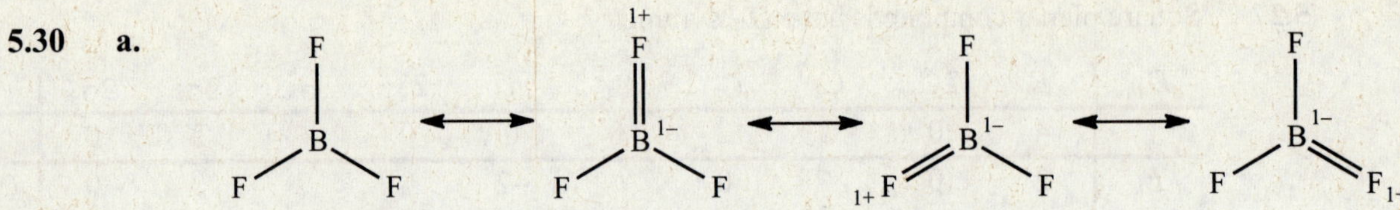

b. The $1a_2''$ orbital near the middle of the figure is the π-bonding orbital.

c. The LUMO, $2a_2''$, is the best orbital for accepting a lone pair.

d. The $1a_2''$ orbital is formed by adding all the p_z orbitals together. The $2a_2''$ orbital is formed by adding the B p_z orbital and subtracting the three F p_z orbitals.

5.31 **a.**

	Ignoring Orbital Lobe Signs	Including Orbital Lobe Signs
$1a_2''$	D_{3h}	C_{3v}
$2a_2''$	D_{3h}	C_{3v}
$1a_2'$	D_{3h}	C_{3h}
$1e''$	C_{2v}	C_1

b. Results should be similar to Figure 5.33. The energies of some of the orbitals in the middle of the diagram are similar, and the order may vary with different calculation methods. In addition, the locations of the nodes in degenerate orbitals (e'' and e') may vary depending on how the software assigns orientations of atomic orbitals. If nodes cut through atomic nuclei, $1e''$ orbitals may have C_2 symmetry, matching the symmetry of the first E'' group orbital shown in Figure 5.32.

c. The table of orbital contributions for each of the orbitals should show the same orbitals as in Figure 5.33. There may be some differences in contributions with different calculation methods, but they should be minor. Assignments to p_x, p_y, and p_z will also differ, depending on how the software defines orientations of orbitals. Semi-empirical calculation AM1 gives these as the major contributors to the specified orbitals:

	$3a_1'$	$4a_1'$	$1a_2''$	$1a_2'$	$2a_2''$
B	$2s$	$2s$	$2p_z$		$2p_z$
F	$2s$	$2s$, $2p_y$	$2p_z$	$2p_x$	$2p_z$

d. Shapes are likely to vary depending on the software used.

5.32 **a.** The shapes of the orbitals, generated using one of the simplest computational methods, Extended Hückel Theory, are shown below, with the most electronegative element shown at right in the heteronuclear cases.

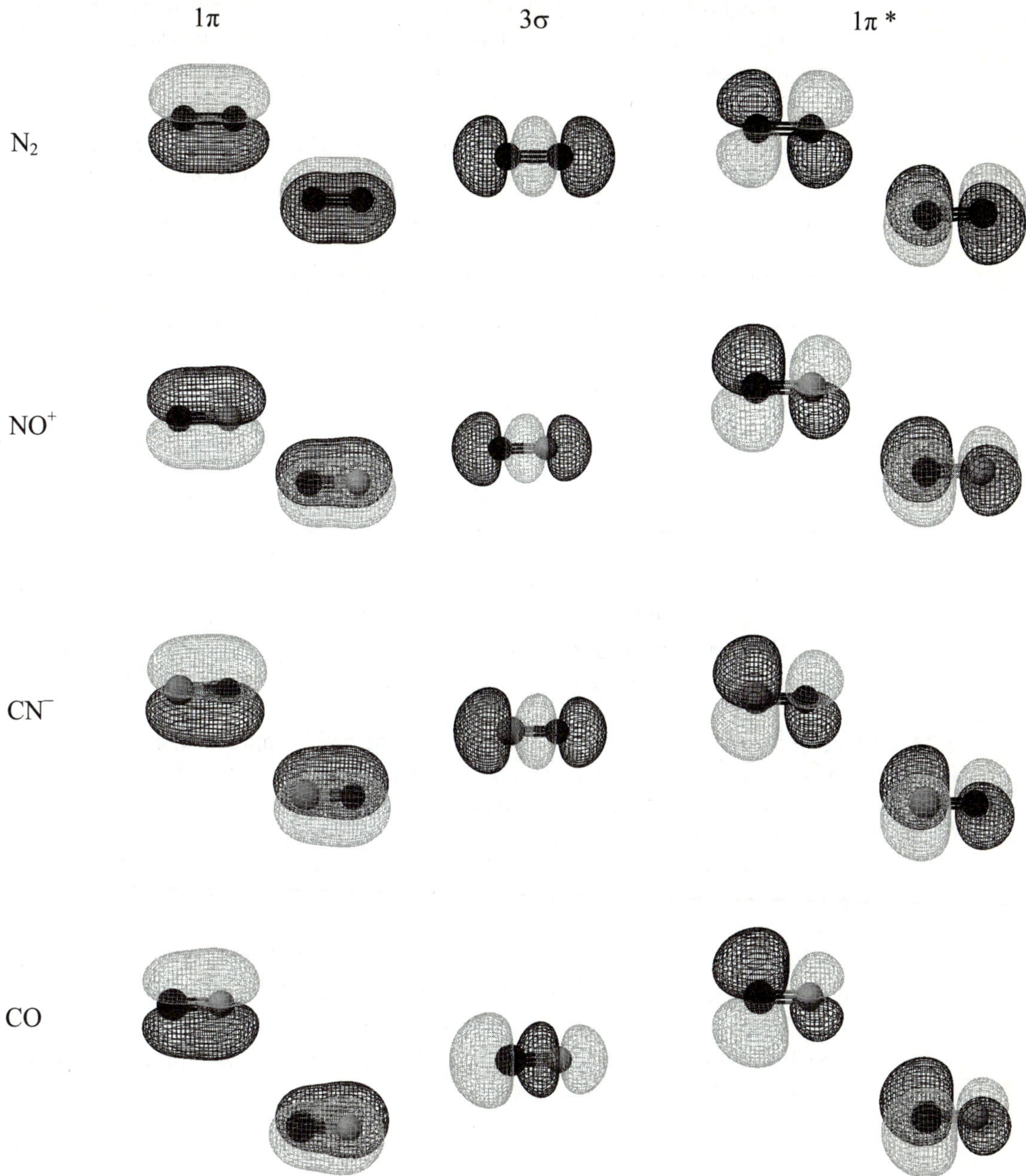

b. In the 1π orbitals (bonding), the lobes are increasingly concentrated on the more electronegative atom as the difference in electronegativity between the two atoms increases. This effect is seen most significantly in CO, where the difference in electronegativity is the greatest.

In the 1π* orbitals, the antibonding partners of the 1π orbitals, the effect is reversed, with the largest lobes now concentrated on the *less* electronegative atoms. The greatest effect is again shown in CO, with the lobes on carbon much larger than those on oxygen.

The 3σ orbitals also show the influence of electronegativity, this time with the lobe extending to the left from the less electronegative atom being the largest, with CO once more showing the greatest effect. This can be viewed in part as a consequence of the 3 σ orbital being a better match for the energy of the less electronegative atom's 2*s* orbital which, together with the $2p_z$ orbital of the same atom, interacts with the $2p_z$ orbital of the more electronegative atom (the atom shown on the right).

c. The results vary greatly depending on the software used. The results using one approach, AM1, are shown below (numerical values are energies in electron volts).

	σ*	π*	σ	π	σ*	σ
		LUMO	HOMO			
CN^-	14.7	0.13	–3.13	–5.10	–9.37	–28.0
CO	5.28	0.94	–13.31	–16.30	–22.00	–41.2
N_2	6.03	1.00	–14.32	–16.19	–21.43	–41.39
NO^+	–4.42	–9.62	–26.13	–28.80	–35.80	–56.89

In this table the energies decrease as the atomic numbers increase (with CO and N_2 giving mixed results). There is considerable mixing of the σ orbitals, a phenomenon that may raise the energy of the σ (HOMO) orbital above the energy of the π orbitals– as is the case in each of these examples.

5.33 Among the trends that should be observed is the effect on the shapes of the π and π* orbitals (see orbitals of CO labeled as 1π and 1π* in Figure 5.14) as the difference in electronegativity between the atoms increases (this trend is also observed in problem 32). For BF and BeNe the lobes of the π orbitals should become increasingly concentrated on the more electronegative atoms, and the lobes of the π* orbitals should become increasingly concentrated on the less electronegative atoms (a pattern that has begun with CO, if the orbital shapes for CO are compared with those of the isoelectronic N_2).

An additional effect is that the size of the protruding orbital lobe of the less electronegative atom should increase as the difference in electronegativity between the atoms increases; this can be see in the 3σ orbital of CO in Figure 5.14 (and in problem 32). Additional trends in the other molecular orbitals can also be noted.

5.34 In one bonding orbital, the H *s* orbitals have the same sign and add to the Be *s* orbital in the HOMO–1 orbital. Subtracting the Be *s* orbital results in the antibonding LUMO. The difference between the two H *s* orbitals added to the Be p_z orbital results in the HOMO; subtracting the Be p_z results in the LUMO+3 orbital. LUMO+1 and LUMO+2 are the Be p_x and p_y orbitals and are nonbonding (and degenerate) in BeH_2. For an energy level diagram, see the solution to Exercise 5.8 in Appendix A.

5.35 BeF_2 is similar to BeH_2, with the addition of π and π* orbitals from the p_x and p_y orbitals, extending over all three atoms. The F p_x orbitals with opposite signs do not combine with the Be orbitals, and neither do the p_y orbitals; the p_x and p_y orbitals form the HOMO and HOMO+1 pair. The answer to Problem 5.15 shows more details.

5.36 The azide orbitals are similar to the CO_2 orbitals, with some differences in contributions from the atomic orbitals because the CO_2 atomic orbitals do not have the identical energies as do the nitrogen atoms. The two highest occupied orbitals of CO_2, BeF_2, and N_3^- all consist of p_x or p_y orbitals of the outer atoms with opposite signs, essentially nonbonding orbitals. The third orbital down has more s orbital contribution from the outer atoms than either of the other two; in those cases, the lower orbital energies of the atoms reduce that contribution. See also the solution to Exercise 5.7 in Appendix A.

5.37 One aspect of ozone's molecular orbitals that should be noted is its π system. For reference, it is useful to compare the bonding π orbital that extends over all three atoms (the atomic orbitals that are involved are shown as molecular orbital 6 in the solution to problem 5.18); this orbital is the lowest in energy of the 3-orbital bonding/nonbonding/antibonding set (orbitals 6, 9, and 10 in problem 5.18) involving the $2p$ orbitals that are not involved in σ bonding. Another bonding/nonbonding/antibonding set can be seen in the molecular orbitals derived from $2s$ orbitals (orbitals 1, 2, and 3 in problem 5.18).

5.38 **a.**

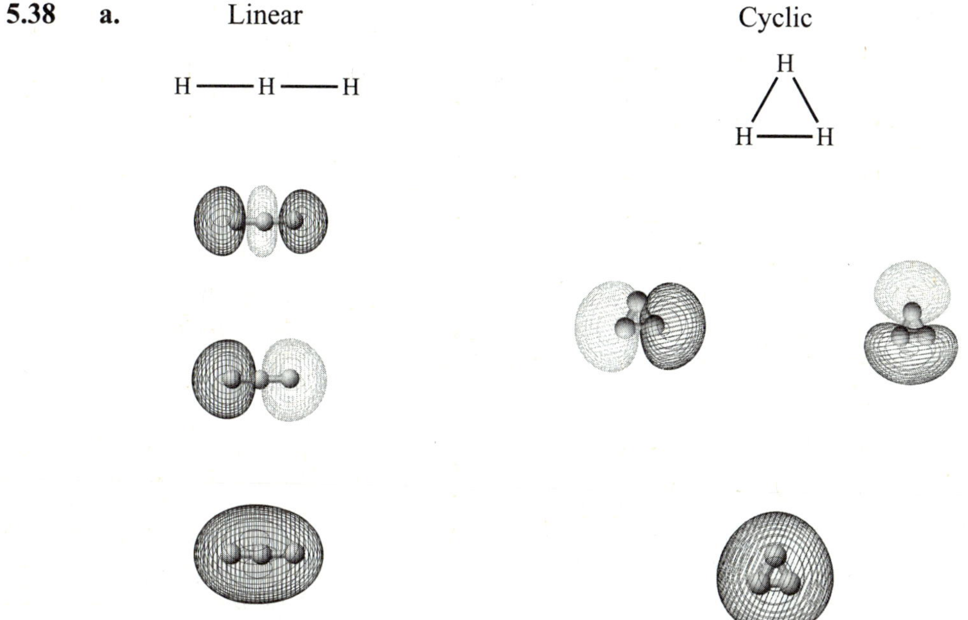

In the linear arrangement, the molecular orbitals shown, from bottom to top, are bonding, nonbonding, and antibonding, with only the bonding orbital occupied. In the cyclic geometry, the lowest energy orbital is bonding, and the other two orbitals are degenerate, each with a node slicing through the center; again, only the lowest energy orbital is occupied.

b. Cyclic H_3^+ is slightly more stable than linear H_3^+, based on the energy of the lowest orbital in an AM1 calculation (–28.4 eV versus –26.7 eV).

5.39 **a.** The full group theory treatment (D_{2h} symmetry), shown on pages 271-273, uses the two bridging hydrogens as one set for group orbitals and the four terminal hydrogens as another set; these sets are shown in Figure 8.11. The representations for these sets can be reduced as follows:

The bridging hydrogens have $\Gamma = A_g + B_{3u}$.

The boron *s* orbitals have $\Gamma = A_g + B_{1u}$.

The p_x orbitals (in the plane of the bridging hydrogens) have $\Gamma = B_{2g} + B_{3u}$.

The p_z orbitals (perpendicular to the plane of the bridging hydrogens) have $\Gamma = A_g + B_{1u}$.

The boron A_g and B_{3u} orbitals combine with the bridging hydrogen orbitals, resulting in two bonding and two antibonding orbitals. Electron pairs in each of the bonding orbitals result in two bonds holding the molecule together through hydrogen bridges.

b. Examples of diborane molecular orbitals are in Figure 8.14, p. 273.

CHAPTER 6: ACID-BASE AND DONOR-ACCEPTOR CHEMISTRY

6.1

	Acid	Base	Definition
a.	BF_3	ClF	Lewis, solvent system
b.	$HClO_4$	CH_3CN	Lewis, Brønsted-Lowry, solvent system
c.	ICl	PCl_5	Lewis, solvent system
d.	ClF_3	NOF	Lewis, solvent system
e.	SO_2	ClO_3^-	Lewis
f.	Pt	XeF_4	Lewis

6.2

	Acid	Base	Definition
a.	XeO_3	OH^-	Lewis
b.	SbF_5	HF	Lewis, solvent system
c.	Sn	NOCl	Lewis
d.	PtF_5	ClF_3	Lewis, solvent system
e.	CH_3COOH	$(benzyl)_3N$	Lewis, Brønsted-Lowry
f.	H_2O	BH_4^-	Lewis

6.3 Al^{3+} is acidic: $[Al(H_2O)_6]^{3+} + H_2O \rightleftharpoons [Al(H_2O)_5(OH)]^{2+} + H_3O^+$

The hydronium ions react with the basic bicarbonate to form CO_2:

$$H_3O^+ + HCO_3^- \rightarrow 2\ H_2O + CO_2\uparrow$$

With pK_a values of 5.0 for $[Al(H_2O)_6]^{3+}$, 6.4 for H_2CO_3, and 2.0 for HSO_4^-, the pH is about 3, low enough to convert the bicarbonate to CO_2.

6.4 An increase in conductivity suggests that ions are formed:

$$BrF_3 + AgF \rightleftharpoons BrF_4^- + Ag^+$$

$$BrF_3 + SnF_4 \rightleftharpoons BrF_2^+ + SnF_5^- \quad \text{or} \quad 2\ BrF_3 + SnF_4 \rightleftharpoons 2\ BrF_2^+ + SnF_6^{2-}$$

6.5 **a.** $3\ ICl \rightleftharpoons I_2Cl^+ + ICl_2^-$ (see Greenwood and Earnshaw, *Chemistry of the Elements*, 2nd ed., p. 827).

b. Both solutes increase the concentration of ions:

$NaCl + ICl \rightarrow Na^+ + ICl_2^-$ (NaCl acts as a base)

$AlCl_3 + 2\ ICl \rightarrow I_2Cl^+ + AlCl_4^-$ ($AlCl_3$ acts as an acid)

6.6 $SnCl_4 + 2Cl^- \rightarrow SnCl_6^{2-}$ is the primary reaction. NH_4Cl in ICl forms NH_4^+ and ICl_2^-, and the chloride ions are then transferred to $SnCl_4$.

6.7 $KF + IF_5 \rightleftharpoons K^+ + IF_6^-$, and the ions conduct electricity.

6.8 **a.** The ions are $[BrF_6]^-$ and $[BrF_4]^+$: $2\ Cs[BrF_6] + [BrF_4][Sb_2F_{11}] \longrightarrow 3\ BrF_5 + 2\ CsSbF_6$

b. $[BrF_6]^-$: O_h (may be distorted) $[BrF_4]^+$: C_{2v}

c. $[BrF_4]^+$ acts as an acid, accepting F^- (Lewis and solvent system concepts).

6.9 $2\ H_2SO_4 \rightleftharpoons H_3SO_4^+ + HSO_4^-$ and $2\ H_3PO_4 \rightleftharpoons H_4PO_4^+ + H_2PO_4^-$ form enough ions to allow conductance in the pure acids.

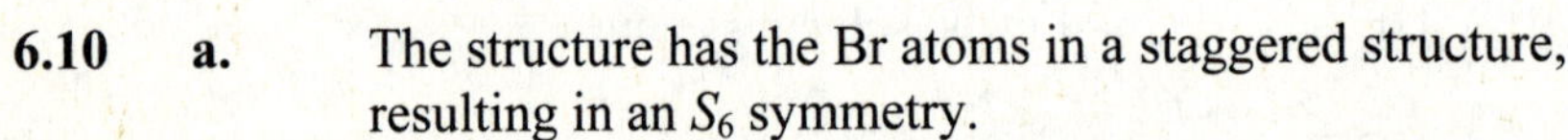

6.10 **a.** The structure has the Br atoms in a staggered structure, resulting in an S_6 symmetry.

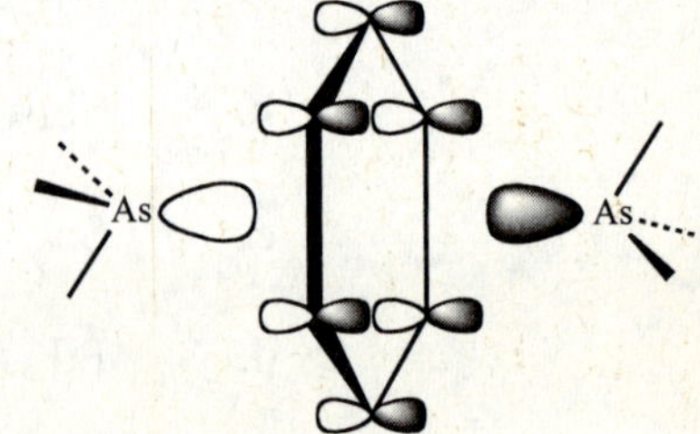

b. It may be easier to visualize this by using tetrahedral As. An sp^3 orbital on each As points inward toward the benzene ring. If the hybrid lobes have opposite signs of their wave functions, they fit the symmetry of the π orbitals of the benzene ring to form bonding and antibonding molecular orbitals. The bonding interaction is shown.

6.11 The very high electronegativity of O in comparison with Al pulls the bonding pair very close to O. This increases the repulsion between the bonding pairs and causes the large angle.

6.12 **a.** The methyl groups in $(CH_3)_3N—SO_3$ donate electrons to the nitrogen, making $(CH_3)_3N$ a stronger Lewis base and strengthening and shortening the N—S bond.

	$(CH_3)_3N—SO_3$	$H_3N—SO_3$
N—S	191.2 pm	195.7 pm
N—S—O	100.1°	97.6°

b. The greater concentration of electrons in the N—S bond of $(CH_3)_3N—SO_3$ increases electron-electron (*bp-bp*) repulsions, opening up the N—S—O bond in comparison with $H_3N—SO_3$.

6.13 **a.** The polarity of the Xe–F bonds concentrates electrons on the fluorine atoms, which act as the centers of Lewis basicity. As shown in the reference, which provides the crystal structure of $[Cd(XeF_2)](BF_4)_2$, the Xe–F–Cd bond is strongly bent at the fluorine and the structure around xenon is nearly linear, with an F–Xe–F bond angle of 179.1°.

b. The BF_4^- ion is smaller than AsF_6^-, and the charge per fluorine is also greater in BF_4^-, making BF_4^- the stronger Lewis base. In addition, the higher oxidation state of cadmium(II) in $[Cd(XeF_2)](BF_4)_2$ enables stronger interaction with the fluorines in BF_4^- than occurs between the silver(I) ion and AsF_6^- in $[Ag(XeF_2)]AsF_6$.

6.14 NO^- is isoelectronic with O_2 and has the electronic structure $\sigma^2\,\pi^2\,\pi^2\,\pi^{*1}\,\pi^{*1}$. Bonding with H^+ depends on which end of the π^* orbital carries more electron density. Calculation shows slightly more electron density on N, making HNO the more likely (bent) molecule.

6.15 **a.** This is similar to the effects described in Section 6.2.7 for I_2. Br_2 forms charge-transfer complexes with donor solvents such as methanol.

b. The 500 nm band ($\pi^* \rightarrow \sigma^*$) should shift to shorter wavelength (higher energy) because the difference in energy between the π^* and σ^* orbitals is greater in $Br_2{\bullet}CH_3OH$ than in Br_2.

6.16 $AlF_3 + F^- \rightarrow AlF_4^-$. The Na^+, AlF_4^- salt can dissolve in HF. When BF_3 is added, it has a stronger attraction for the fluoride ions, with the reaction $AlF_4^- + BF_3 \rightarrow AlF_3\downarrow + BF_4^-$.

6.17 Soft metal ions do not combine with oxygen as strongly as hard metal ions, so reactions like

$$2\,HgO \xrightarrow{\Delta} 2\,Hg + O_2\uparrow \qquad 2\,CuO \xrightarrow{\Delta} 2\,Cu + O_2\uparrow$$

$$2\,Ag_2O \xrightarrow{\Delta} 4\,Ag + O_2\uparrow$$

$$CuO + C \xrightarrow{\Delta} Cu + CO\uparrow \qquad 2\,CuO + C \xrightarrow{\Delta} 2\,Cu + CO_2\uparrow$$

are more easily carried out. Reduction of some of the softer metals can be carried out with relatively low temperatures (campfires); some think use of rocks containing ores could have led to accidental reduction to the metals and discovery of the smelting process. Harder metals such as iron require much higher temperatures for the reduction process.

6.18 Hg is a much softer metal, and combines with the soft sulfide ion more strongly. Zinc and cadmium are more borderline metals, and combine with all the anions with more nearly equal facility.

6.19 When any of these salts vaporize, the vapor phase consists of molecules. When they are in the solid state, they are ionic, with some covalent properties. The liquid state is between the two, and can be made up of either ions, covalent molecules, or something between these two extremes. If the liquid is molecular, vaporization should be easier (molecules in the liquid phase being converted to molecules in the vapor phase). If the liquid is mostly ionic, vaporization requires greater molecular change and should be more difficult. Using this criterion, the most ionic liquids should be ZnF_2 and CdF_2 and the most molecular liquids should be HgF_2 and $HgCl_2$. On a more general view, mercury as the softest metal in the series forms the more molecular compounds and zinc as the hardest forms the more ionic compounds. Cadmium forms the most molecular compound with the borderline bromide.

6.20 **a.** pyridine + BF_3:

$$\Delta H = -\left(E_{py}E_{BF_3} + C_{py}C_{BF_3}\right)$$
$$= -[(1.17)(9.88) + (6.40)(1.62)] = -21.9 \text{ kcal/mol or } -91.6 \text{ kJ/mol, about 10\% low}$$

pyridine + $B(CH_3)_3$:

$$\Delta H = -\left(E_{py}E_{B(CH_3)_3} + C_{py}C_{B(CH_3)_3}\right)$$
$$= -[(1.17)(6.14) + (6.40)(1.70)] = -18.1 \text{ kcal/mol or } -75.7 \text{ kJ/mol, 5 to 18\% high}$$

b. Fluorine is electron-withdrawing, CH_3 electron-releasing. Therefore, B in BF_3 carries a greater positive charge and interacts more strongly with Lewis bases such as pyridine.

c. The harder acid BF_3 interacts more strongly with the borderline base pyridine.

6.21 $NH_3 + BF_3$:

$$\Delta H = -\left(E_{NH_3}E_{BF_3} + C_{NH_3}C_{BF_3}\right)$$
$$= -[(1.36)(9.88) + (3.46)(1.62)] = -19.0 \text{ kcal/mol or } -79.5 \text{ kJ/mol}$$

$NH_3 + B(CH_3)_3$:

$$\Delta H = -\left(E_{NH_3}E_{B(CH_3)_3} + C_{NH_3}C_{B(CH_3)_3}\right)$$
$$= [(1.36)(6.14) + (3.46)(1.70)] = -14.2 \text{ kcal/mol or } -59.4 \text{ kJ/mol}$$

The order is pyridine + BF_3 > NH_3 + BF_3 > pyridine + $B(CH_3)_3$ > NH_3 + $B(CH_3)_3$. The change From BF_3 to $B(CH_3)_3$ is larger than the change from pyridine to NH_3.

6.22 Absolute hardness parameters:

	I	A	η
BF_3	15.81	–3.5	9.7
NH_3	10.7	–5.6	8.2
C_5H_5N	9.3	–0.6	5.0
$N(CH_3)_3$	7.8	–4.8	6.3

The difference between the HOMO of pyridine (9.3) and the LUMO of BF_3 (–3.5) is smaller than the other possible interaction, so this pair has the largest $-\Delta H$, in spite of the difference in hardness. By comparison with the nitrogen compounds, $B(CH_3)_3$ is expected to have an absolute hardness lower than that of BF_3, approximately 7.5-8.

6.23 CsI and LiF fit Basalo's principle that ions of similar size and equal (but opposite) charge form the least soluble salts. CsF and LiI have ionic sizes that are very different, and they do not fit as well into an ionic lattice. In addition, CsI and LiF are soft-soft and hard-hard combinations, which combine better than the hard-soft LiI and soft-hard CsF.

6.24 $H_3C{-}CH_3 \cdots H{-}O{-}H \longrightarrow H_3C{-}O{-}H + CH_3{-}H$ is unlikely. C is soft, O is hard.

$H_3C{-}C({=}O){-}CH_3 \cdots H{-}O{-}H \longrightarrow H_3C{-}C({=}O){-}O{-}H + CH_3{-}H$ is more likely. Adding the carbonyl oxygen makes C harder, and C in CH_3 and the H atom are soft.

6.25 **a.** Solubilities: $MgSO_4 > CaSO_4 > SrSO_4 > BaSO_4$
Electrostatic forces predict the reverse order due to cation sizes, but the larger cations fit better with the large sulfate anion in the crystals. Hydration of the cations is the strongest for Mg^{2+}, weakest for Ba^{2+}, agreeing with the solubility order.

b. Solubilities: $PbCl_2 > PbBr_2 > PbI_2 > PbS$
As a moderately soft cation, Pb^{2+} has stronger interactions with the softer anions (order $Cl^- > Br^- > I^- > S^{2-}$). In addition, hydration of the anions is largest for the chloride, smallest for sulfide, based on size.

6.26 The order is Al–OC–W (see Figure 13.17). The oxygen of CO is the "harder" end of this molecule and bonds with the harder Al atom; carbon and W engage in a soft–soft interaction.

6.27 **a.** TeH_2 is the strongest acid, because Te is more electronegative than either Sn or Sb. Therefore, the hydrogen is more positive and acidic.

b. NH_3 is the strongest base because N is more electronegative than either P or Sb. The charge density is large enough that it not only holds its own hydrogens, but attracts another.

c. $(CH_3)_3N$ is the strongest base in the gas phase because the methyl groups contribute electron density to the nitrogen. In solution, the order is scrambled, probably due to solvation (pp. 213).

d. 4-Methylpyridine > pyridine > 2-methylpyridine. Again, the methyl group adds electron density to the N. However, with methyl in the 2 position, steric hindrance makes bonding to BMe_3 more difficult.

6.28 In general, oxide ion reacts with water to form hydroxide: $O^{2-} + H_2O \rightarrow 2\ OH^-$ unless other factors prevent it. In B_2O_3, the small, hard B^{3+} holds on to the oxide ions strongly. As a result, $B_2O_3 + 3\ H_2O \rightarrow 2B(OH)_3 \rightleftharpoons H^+ + H_2BO_3^-$, and the solution is very weakly acidic ($pK_a = 9.25$). In Al_2O_3, the Al^{3+} ion is larger and softer. It can form either $[Al(OH)_4]^-$ (acting as an acid) or $[Al(H_2O)_6]^{3+}$ (acting as a base), depending on the other species in solution. Sc^{3+} is still larger and softer, so it combines better with water than with hydroxide ion. As a result, the reaction $Sc_2O_3 + 15\ H_2O \rightarrow 2[Sc(H_2O)_6]^{3+} + 6\ OH^-$ is possible.

6.29 **a.** $CaH_2 + 2\ H_2O \rightarrow Ca^{2+} + 2\ H_2\uparrow + 2\ OH^-$ Calcium has a lower electronegativity than hydrogen, so CaH_2 is $Ca^{2+}(H^-)_2$ and the hydride ions react with the positive hydrogens of water.

b. $HBr + H_2O \rightarrow H_3O^+ + Br^-$ Bromine is more electronegativite than hydrogen, so the hydrogen is strongly positive and is readily lost to the lone pair of water.

c. $H_2S + H_2O \rightleftharpoons H_3O^+ + HS^-$ Sulfur is slightly more electronegative than hydrogen, and the positive hydrogen in H_2S can dissociate to a small extent.

d. $CH_4 + H_2O \rightarrow$ no reaction. The C–H bond is almost nonpolar; the hydrogens are not positive enough to be attracted to the water lone pair.

6.30 $BF_3 > B(CH_3)_3 > B(C_2H_5)_3 > B(C_6H_2(CH_3)_3)_3$ Alkyl groups are electron-donating and increase the electron density on B and reduce the attraction for the lone pair of NH_3. The large mesityl groups reduce the adduct formation because they are too bulky to fold back readily into the required tetrahedral geometry.

6.31 **a.** CH_3NH_2 is a stronger base. The methyl group pushes electron density onto the nitrogen.

b. Although 2-methylpyridine is the stronger base with smaller acid molecules, the methyl group interferes with adduct formation with trimethylboron (F-strain) and the pyridine-trimethylboron formation is stronger.

c. Trimethylboron forms a stronger adduct with ammonia because the three phenyl rings of triphenylboron cannot bend back readily to allow the boron to become tetrahedral (B-strain).

6.32 **a.** With the acids listed in order of increasing acidity:

	H_3AsO_4	H_2SO_3	H_2SO_4	$HMnO_4$
pK_a (9-7n)	2	2	–5	–12
pK_a (8-5n)	3	3	–2	–7
pK_a (exp)	9.2	2.2	1.8	–11

b. With the acids listed in order of increasing acidity:

	$HClO$	$HClO_2$	$HClO_3$	$HClO_4$
pK_a (9-7n)	9	2	–5	–12
pK_a (8-5n)	8	3	–2	–7
pK_a (exp)	7.4	2	–1	–10

6.33 Dimethylamine acts as weak base in water, with a very small amount of OH^- provided by the reaction $(CH_3)_2NH + H_2O \rightleftharpoons (CH_3)_2NH_2^+ + OH^-$. Acetic acid is a stronger acid than water, so dimethylamine acts as a stronger base, and the reaction

$$(CH_3)_2NH + HOAc \rightarrow (CH_3)_2NH_2^+ + OAc^-$$

goes to completion. 2-Butanone is a neutral solvent; there is no significant acid-base reaction with dimethylamine.

6.34 SbF_5 in HF reacts to increase the H^+ concentration and decrease H_0:

$SbF_5 + HF \rightleftharpoons H^+ + SbF_6^-$ These ions then can react with alkenes.

6.35 **a.** As the Lewis acids BF_3 and BCl_3 interact with NH_3, the geometry around boron changes from planar to trigonal pyramidal; however, in accord with the LCP model the nonbonded F···F and Cl···Cl distances are nearly constant, suggesting that these atoms remain in contact with each other. Because these nonbonded distances remain essentially constant, the boron–halogen distance must increase as the distortion from trigonal geometry occurs. Because of the strength of the B–F bond, a consequence of the large electronegativity difference between B and F and the small size of the fluorine atom, more energy is required to distort BF_3 from planarity than to similarly distort BCl_3. (Calculations in support of this argument are presented in the reference.) The consequence of this energy requirement is that BF_3 is a weaker Lewis base than BCl_3 toward NH_3. The article does not address the relative Lewis basicity of BBr_3, but a similar argument could apply for this compound.

b. This article does not consider the LCP model but focuses on ab initio calculations on the adducts $X_3B–NH_3$. These calculations show a higher B–N bond dissociation energy in the BCl_3 adduct than in the BF_3 adduct. Although many factors are involved, the principal factor in the stronger bond in the BCl_3 adduct is that BCl_3 has a lower energy LUMO that is able to addict more strongly with the donor orbital of NH_3, giving a stronger covalent interaction (and stronger B–N bond) in $Cl_3B–NH_3$.

6.36 **a.** The N_2O acts as a bridge between the phosphorus atom of the phosphine and the boron, with the more electronegative oxygen forming a bond with the Lewis acid $B(C_6F_5)_3$.

b.

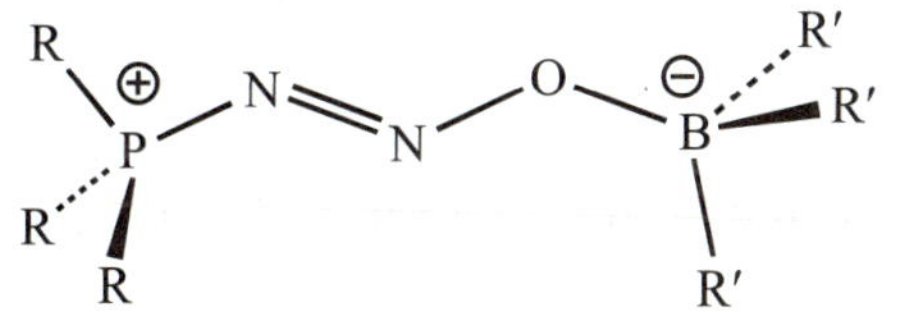

R ⊕ P N N O ⊖ B R′ R′ R′ R R

$R = t\text{-}C_4H_9$

$R' = C_6F_5$

6.37 A frustrated Lewis pair is described briefly in problem 6.36. See the reference for a more detailed definition and applications of FLPs.

6.38 **a.** The energy diagram and the orbitals are shown below.

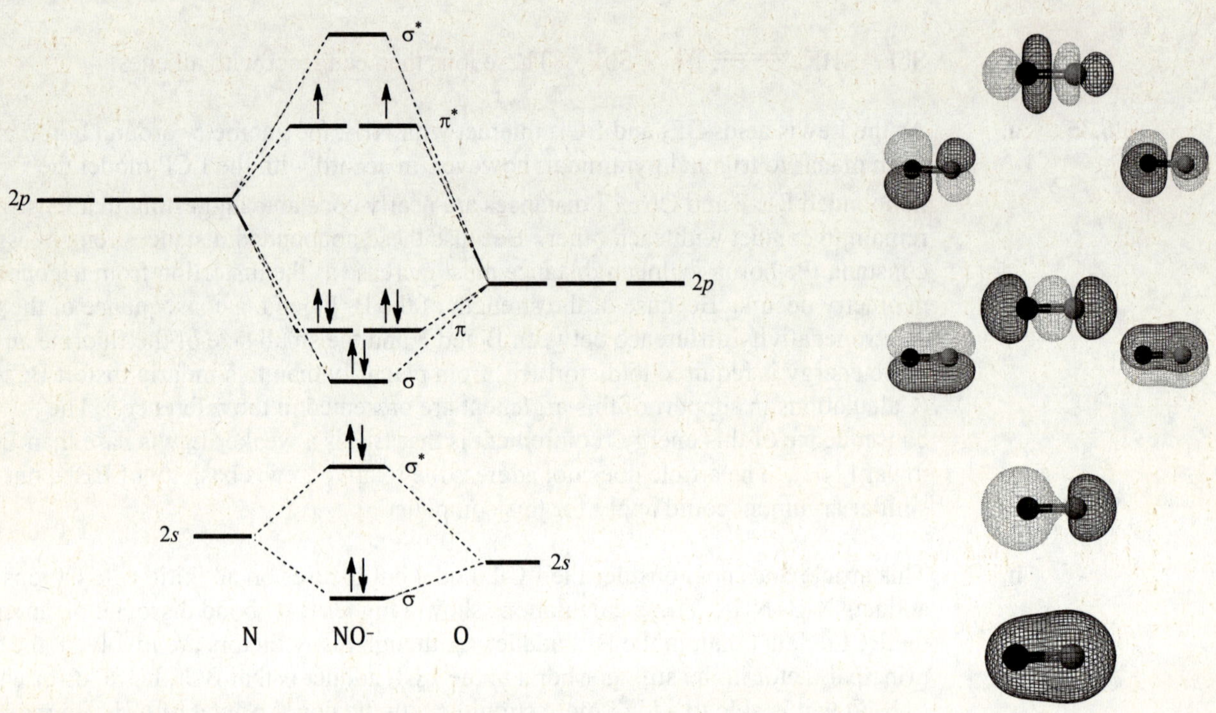

The HOMO is the π*, with greater concentration of electron density on N.

b. The π* HOMO, concentrated primarily on the nitrogen of NO^-, can combine with the empty H^+ 1*s* orbital, forming HNO, a bent molecule. Comparing HNO and NOH:

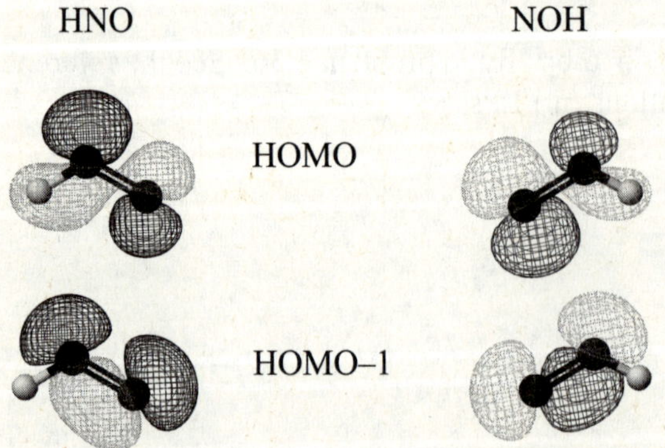

Neither is highly stable, but calculations predict that HNO is more stable than NOH.

6.39

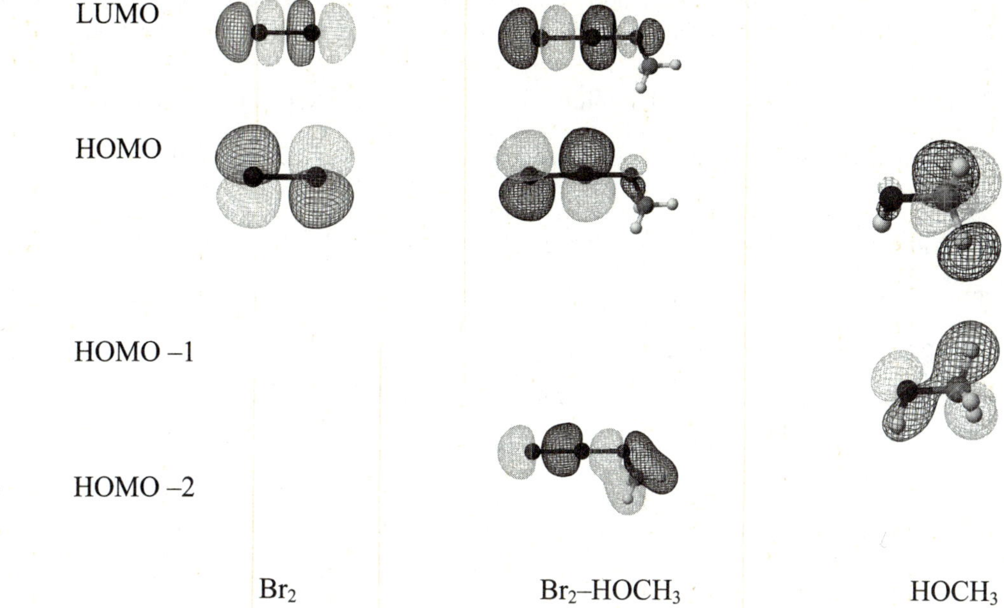

The interaction of the HOMO of the methanol and the LUMO of the Br_2 results in the LUMO and HOMO –2 orbitals of the adduct. The geometry shown has the Br_2 at approximately a trigonal angle (H—O—Br is 113° and C—O—Br is 106° in a PM3 calculation).

6.40 **a.** For orbitals of BF_3 and NH_3, see Figures 5.33 and 5.31, respectively.

b. The B–N bonding molecular orbital, shown below, is polarized toward the more electronegative nitrogen; the matching antibonding orbital, which has a node between the boron and nitrogen atoms, is polarized toward the boron.

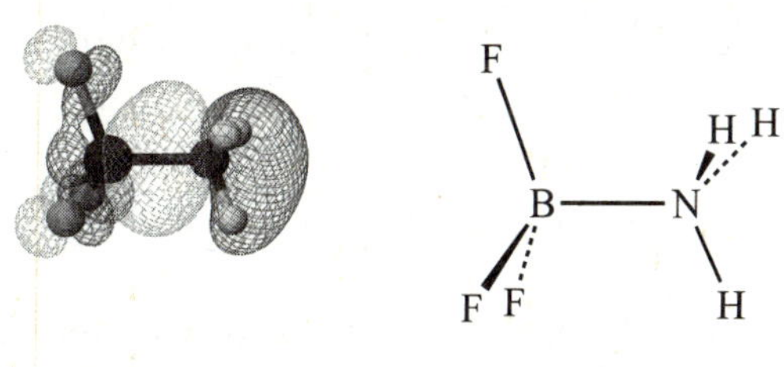

CHAPTER 7: THE CRYSTALLINE SOLID STATE

7.1 **a.** O_h **b.** D_{4h} **c.** O_h **d.** T_d

e. The image in Figure 7.10, which has D_{3d} symmetry, actually consists of three unit cells. For an image of a single unit cell, which has point group C_{2h}, see page 556 in the Greenwood and Earnshaw reference in the "General References" section.

7.2 The unit cell dimension is $2r$, the volume is $8r^3$. Since this cell contains one molecule whose volume is $4/3\ \pi\ r^3$, the fraction occupied is $\frac{4/3\pi r^3}{8r^3} = 0.524 = 52.4\%$

7.3 The unit cell length for a primitive cubic cell is $2r$. Using the Pythagorean theorem, we can calculate the face diagonal as $\sqrt{(2r)^2 + (2r)^2} = 2.828r$ and the body diagonal as $\sqrt{(2.828r)^2 + (2r)^2} = 3.464r$. Each corner atom contributes r to this distance, so the diameter of the body center is $1.464r$ and the radius is $0.732r$, 73.2% of the corner atom size.

7.4 **a.** A face-centered cubic cell is shown preceding Exercise 7.1 in section 7.1.1. The cell contains $6 \times \frac{1}{2} = 3$ atoms at the centers of the faces and $8 \times \frac{1}{8} = 1$ atom at the corners, a total of 4 atoms. The total volume of these 4 atoms $= 4 \times \frac{4}{3}\pi r^3 = \frac{16}{3}\pi r^3$, where r is the radius of each atom (treated as a sphere).

As can be seen in the diagram, the diagonal of a face of the cube $= 4r$. By the Pythagorean theorem, the dimension (length of a side) of the cube $= \frac{\text{diagonal}}{\sqrt{2}} = \frac{4r}{\sqrt{2}}$.

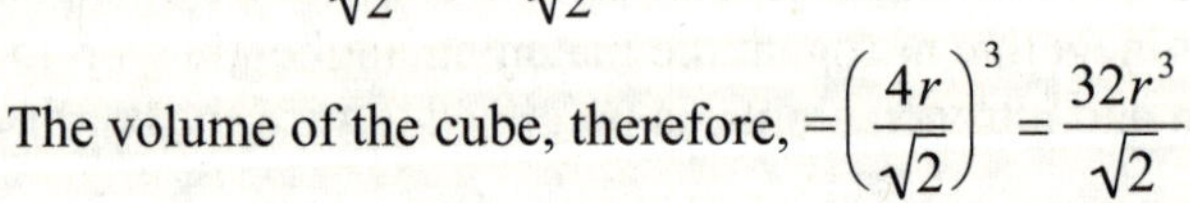

The volume of the cube, therefore, $= \left(\frac{4r}{\sqrt{2}}\right)^3 = \frac{32r^3}{\sqrt{2}}$

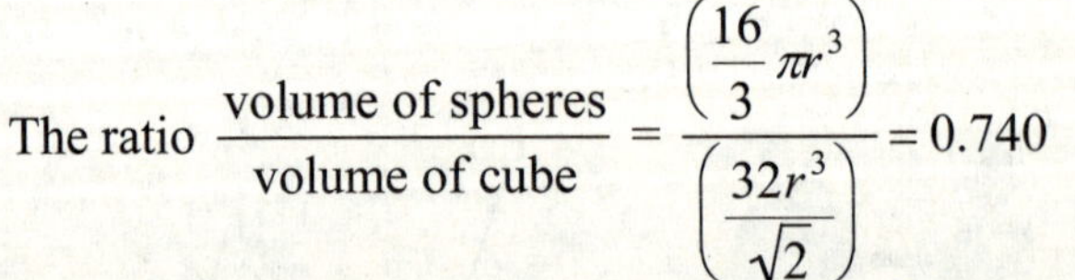

The ratio $\frac{\text{volume of spheres}}{\text{volume of cube}} = \frac{\left(\frac{16}{3}\pi r^3\right)}{\left(\frac{32r^3}{\sqrt{2}}\right)} = 0.740$

Therefore, 74.0% of the volume of the cube is occupied by the spheres.

b. In a body-centered cube the unit cell contains one atom in the center of the cube and $8 \times \frac{1}{8} = 1$ atom at the corners, a total of 2 atoms. The total volume of these 2 atoms $= 2 \times \frac{4}{3}\pi r^3 = \frac{8}{3}\pi r^3$, where r is the radius of each atom.

Because atoms touch along the diagonal, the diagonal of the cube $= 4r$. Using the Pythagorean theorem, it can be shown that the dimension of the cube $= \frac{\text{diagonal}}{\sqrt{3}} = \frac{4r}{\sqrt{3}}$.

Therefore, the volume of the cube $= \left(\dfrac{4r}{\sqrt{3}}\right)^3 = \dfrac{64r^3}{3\sqrt{3}}$

The ratio $\dfrac{\text{volume of spheres}}{\text{volume of cube}} = \dfrac{\left(\dfrac{8}{3}\pi r^3\right)}{\left(\dfrac{64r^3}{3\sqrt{3}}\right)} = 0.680$ or 68.0%

7.5 In the table below, CaF_2 is considered to have a fluoride ion in the body center of the overall unit cell and calcium ions in the body centers of the subunits (labeled "internal").

Compound	Corners	Edges	Face Centers	Body Centers	Internal	Total	Type
NaCl cations	8 × 1/8		6 × 1/2			4	MX
NaCl anions		12 × 1/4		1 × 1		4	MX
CsCl cations	8 × 1/8					1	MX
CsCl anions				1 × 1		1	MX
CaF_2 cations					4 × 1	4	MX_2
CaF_2 anions	8 × 1/8	12 × 1/4	6 × 1/2	1 × 1		8	MX_2

7.6 LiBr has a formula weight of 86.845, and the unit cell contains four cations and four anions (or four formula units per molecular unit cell).

$$\frac{86.845 \text{ g mol}^{-1}}{3.464 \text{ g cm}^{-3}} = 25.07 \text{ cm}^3\text{mol}^{-1} \times \frac{10^{-6}\text{m}^3}{\text{cm}^3} = 2.507 \times 10^{-5}\text{m}^3\text{mol}^{-1}$$

$$\frac{2.507 \times 10^{-5}\text{m}^3\text{mol}^{-1}}{6.022 \times 10^{23}\text{units mol}^{-1}} \times \frac{4 \text{ units}}{\text{unit cell}} = \frac{1.665 \times 10^{-28}\text{m}^3}{\text{unit cell}}$$

$$\sqrt[3]{1.665 \times 10^{-28}\text{m}^3} = 5.502 \times 10^{-10}\text{m} = \text{unit cell length}$$

$$2(r_+ + r_-) = 5.502 \times 10^{-10}\text{m};\ r_+ + r_- = 2.751 \times 10^{-10}\text{m} = 275.1 \text{ pm}$$

The sum of the ionic radii from Appendix B.1 is 272 pm.

7.7 CsCl has 8 Cl^- at the corners of the unit cell cube, with the Cs^+ at the center.
$r_+/r_- = 188/167 = 1.13$.

CaF_2 has the same structure in a single cube of F^- ions, but only half the cubes contain Ca^{2+}.
$r_+/r_- = 126/119 = 1.06$.

Both should have coordination number = 12 based on the radius ratios.

7.8 Figure 7.8 shows the zinc blende unit cell, which contains four S atoms (net) in an fcc lattice and four Zn atoms in the body centers of the alternate smaller cubes. The diagrams below are each a view of two layers of such a cell, with • indicating a Zn atom in this layer and a o indicating an S atom in the layer below. The next pair of layers either above or below these has the opposite pattern, and the third repeats the original. The fcc lattice pattern can be seen for the Zn atoms (four corners and face center on the top face, four face centers in the middle later, and four corners and face center on the bottom face). Each S atom (and each Zn atom) has two nearest neighbors in the layer above and two in the layer below, in the arrangement for a tetrahedral hole. Extending the patterns below shows the S fcc lattice.

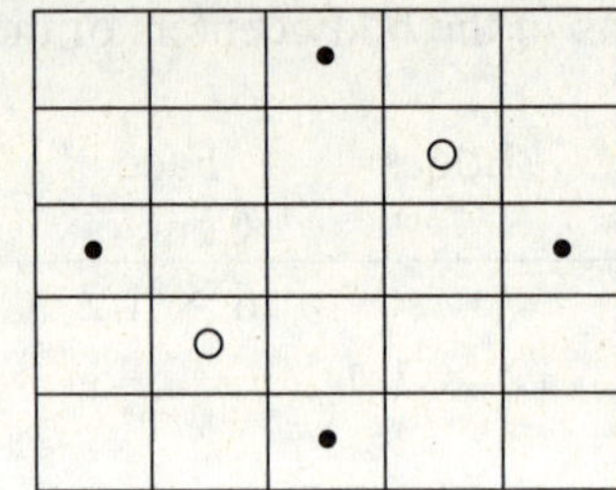

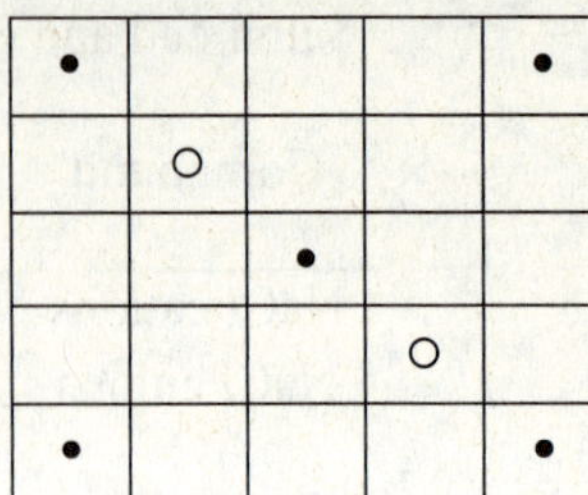

7.9 The graphite layers have essentially the same energy levels as benzene, but each level becomes a wider band because of the large number of atoms. This leads to the energy levels shown at right, with the bands coming from the lowest energy π orbitals filled and those coming from the highest energy π orbitals empty. The difference between the highest occupied and lowest unoccupied bands is small enough to allow conduction electrons to make the jump and the electrons and holes to move within the bands. Conduction perpendicular to the layers is smaller, because there are no direct orbitals connecting them. In polycrystalline graphite, the overall conductance is an average of the two. (The π orbitals of benzene are shown in Figure 13.22, page 511.)

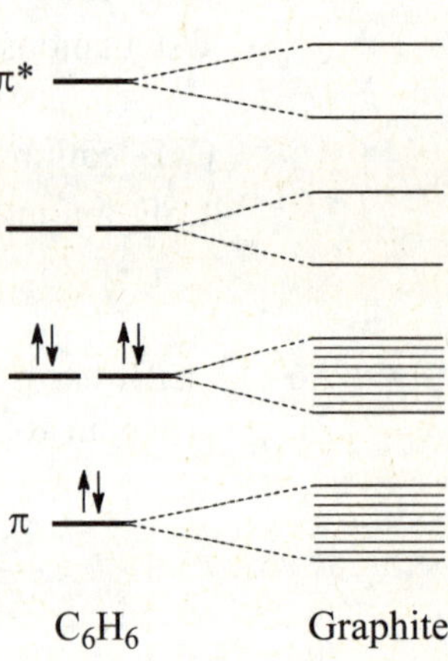

In diamond, each carbon atom has four σ bonds to its nearest neighbors. The gap between these filled orbitals and the corresponding antibonding orbitals is larger, effectively limiting conductance.

7.10 Solutions of alkali halides in water conduct electricity. This does not prove that they are ionic as solids, but is suggestive of ions in the solid state. Their high melting points are also suggestive of ionic structures, and the molten salts also conduct electricity. Perhaps the most conclusive evidence is from X-ray diffraction studies, in which these compounds show uniform cation–anion distances. If they were molecular species, the interatomic distances within a molecule should be smaller than the interatomic distances between molecules.

7.11 Hg(I) appears in compounds such as Hg_2^{2+} units. The $6s^1\ 4f^{14}\ 5d^{10}$ structure of Hg^+ forms σ and σ* molecular orbitals from the *s* atomic orbitals, allowing the two *s* electrons to pair in the σ orbital for a diamagnetic unit.

7.12 **a.** Forming anions from neutral atoms results in addition of an electron. More electrons means a larger size, due to increasing electron–electron repulsion. By the same argument, forming cations from neutral atoms results in removal of an electron and smaller size. The cations have fewer electrons and less electron–electron repulsions.

b. The oxide ion is larger than the fluoride ion because its nuclear charge is smaller; the oxygen nucleus has less attraction for the electrons.

7.13 Radius ratios for the alkali halides:

Ions	Radii (pm)	Li^+	Na^+	K^+	Rb^+	Cs^+
Radii (pm)		90	116	152	166	206
F^-	119	0.76	0.98	1.28	1.40	1.73
Cl^-	167	0.54	0.70	0.91	0.99	1.23
Br^-	182	0.49	0.64	0.84	0.91	1.13
I^-	206	0.44	0.56	0.74	0.81	1.00

Cation–anion combinations with a light outline have radius ratios indicating a coordination number of 8, and those with a dark outline indicate a coordination number of 12. All of these compounds actually have the sodium chloride lattice (CN = 6), so only six of the 20 give the correct predicted structure, with KI also very close to the CN = 6 ratio.

7.14 Cation radius (pm) calculated from interionic distance and anion radius:

	Li	Na	Ag
F (119 pm)	82	112	127
Cl (167 pm)	90	114	110
Br (182 pm)	93	116	106
r_+ (Appendix B.1)	90	116	129

Hard–soft combinations (LiBr, AgF) have larger distances than calculated; hard–hard (LiF) and soft–soft (AgBr) have smaller distances. NaF is a hard–hard combination by this criterion. The radius ratios for NaF and AgF are large enough to predict CN = 8, the CsCl structure. All the others fit the CN = 6, NaCl structure criteria.

7.15

	$\frac{1}{2}\,Cl_2\,(g) \rightarrow Cl\,(g)$	239/2 = 119.5 (Energies in kJ/mol)
	$Cl\,(g) + e^- \rightarrow Cl^-\,(g)$	–EA
	$Na\,(s) \rightarrow Na\,(g)$	109
	$Na\,(g) \rightarrow Na^+\,(g) + e^-$	496 (=5.14 eV × 96.4853 kcal/mol/eV)
	$Na^+\,(g) + Cl^-\,(g) \rightarrow NaCl\,(s)$	–772
Total:	$Na\,(s) + \frac{1}{2}\,Cl_2\,(g) \rightarrow NaCl\,(s)$	–413

EA = 366 kJ/mol

$$U = \frac{NMZ_+Z_-}{r_0}\left[\frac{e^2}{4\pi\varepsilon_0}\right]\left(1-\frac{\rho}{r_0}\right) = \frac{6.022\times10^{23}\times1.74756\times1\times(-1)}{281\times10^{-12}\,\text{m}}\times2.3071\times10^{-28}\,\text{Jm}\times\left(1-\frac{30}{281}\right)$$

$$= -772 \text{ kJ/mol}$$

7.16 CaO has charges of 2+ and 2–, and radii of 114 and 126 pm, total distance of 240 pm.

KF has charges of 1+ and 1–, and radii of 152 and 119 pm, total distance of 271 pm.

The distance in CaO is 13% smaller and the charge factor is four times as large, both leading to stronger interionic attraction and contributing to the hardness of the crystal.

MgO has charges of 2+ and 2–, and radii of 86 and 126 pm, total distance of 212 pm, with NaCl structure and Madelung constant = 1.75.

CaF_2 has charges of 2+ and 1–, and radii of 126 and 119 pm, total distance of 245 pm, with the fluorite structure and Madelung constant = 2.52.

The size difference and charges favor stronger MgO interionic attraction, enough to overcome the Madelung constant difference.

7.17 $MgCl_2$ has a rutile structure, so it will be used for the $NaCl_2$ calculation.

The lattice energy for a rutile structure with combined radii of 270 pm (either $NaCl_2$ or $MgCl_2$) is:

$$U = \frac{NMZ_+Z_-}{r_0}\left[\frac{e^2}{4\pi\varepsilon_0}\right]\left(1-\frac{\rho}{r_0}\right)$$

$$U = \frac{6.022\times10^{23}\,\text{mol}^{-1}\times 2.385\times 2\times(-1)}{270\times10^{-12}\,\text{m}}\times 2.307\times10^{-28}\,\text{Jm}\times\left(1-\frac{30}{270}\right)$$

$$= -2{,}182\ \text{kJ mol}^{-1}$$

The overall energy of formation of $MgCl_2$ is negative. For this calculation, the only differences between $MgCl_2$ and $NaCl_2$ are in the vaporization and ionization of the metal. The big difference is in the second ionization energy. $Na^+ \rightarrow Na^{2+} + e^-$, removing an electron from a closed 2*p* orbital, requires much more energy than the corresponding $Mg^+ \rightarrow Mg^{2+} + e^-$ reaction that removes the second 3*s* electron.

In fact, the two metals differ by about 2,800 kJ/mol in these steps (all in kJ/mol):

$Na\ (s) \rightarrow Na(g)$	107	$Mg\ (s) \rightarrow Mg\ (g)$	147
$Na\ (g) \rightarrow Na^+(g) + e^-$	495	$Mg\ (g) \rightarrow Mg^+(g) + e^-$	738
$Na^+\ (g) \rightarrow Na^{2+}(g) + e^-$	4,562	$Mg^+\ (g) \rightarrow Mg^{2+}(g) + e^-$	1,451
Totals: $Na\ (s) \rightarrow Na^{2+}(g) + 2e^-$	5,164	$Mg\ (s) \rightarrow Mg^{2+}(g) + 2e^-$	2,336

Therefore, it is extremely unlikely that $NaCl_2$ can be made.

Repeating the process for a sodium chloride lattice with the sum of radii 283 pm to compare the lattice energy and the energies necessary to form Na^+ and Mg^+:

$$U = \frac{NMZ_+Z_-}{r_0}\left[\frac{e^2}{4\pi\varepsilon_0}\right]\left(1-\frac{\rho}{r_0}\right)$$

$$U = \frac{6.022\times10^{23}\text{mol}^{-1}\times1.748\times1\times(-1)}{283\times10^{-12}\text{m}}\times2.307\times10^{-28}\,\text{Jm}\times\left(1-\frac{30}{283}\right)$$

$$= -767 \text{ kJ mol}^{-1}$$

Na (s) $\rightarrow$ Na(g)	107	Mg (s) $\rightarrow$ Mg (g)	147
Na (g) $\rightarrow$ Na^+(g) + e^-	495	Mg (g) $\rightarrow$ Mg^+(g) + e^-	738
Na (s) $\rightarrow$ Na^+(g) + e^-	602	Mg (s) $\rightarrow$ Mg^+(g) + e^-	885

In this case, formation of MgCl seems possible, but formation of $MgCl_2$ is so much more favorable (nearly triple the lattice energy) that it is unlikely for the reaction to stop at the MgCl stage.

7.18

½ Br_2 (l) $\rightarrow$ ½ Br_2 (g)	14.9	(Energies in kJ/mol)
½ Br_2 (g) $\rightarrow$ Br (g)	190.2	
Br (g) + e^- $\rightarrow$ Br^-	–324.7	
K (s) $\rightarrow$ K (g)	81.3	
K (g) $\rightarrow$ K^+ + e^-	418.8	
K^+ + Br^- $\rightarrow$ KBr (s)	–661.8	
Total	–281.3 kJ mol^{-1}	for K (s) + ½ Br_2 (l) $\rightarrow$ KBr (s)

For a sodium chloride lattice with total radii of 334 pm:

$$U = \frac{NMZ_+Z_-}{r_0}\left[\frac{e^2}{4\pi\varepsilon_0}\right]\left(1-\frac{\rho}{r_0}\right)$$

$$U = \frac{6.022\times10^{23}\text{mol}^{-1}\times1.748\times1\times(-1)}{334\times10^{-12}\text{m}}\times2.307\times10^{-28}\,\text{Jm}\times\left(1-\frac{30}{334}\right)$$

$$= -661.8 \text{ kJ mol}^{-1}$$

7.19

½ O_2 (g) $\rightarrow$ O (g)	247	(Energies in kJ/mol)
O (g) + 2 e^- $\rightarrow$ O^{2-}	603	
Mg (s) $\rightarrow$ Mg (g)	37	
Mg (g) $\rightarrow$ Mg^{2+} + 2e^-	2188	
Mg^{2+} + O^{2-} $\rightarrow$ MgO (s)	–3934	
Total	–859 kJ mol^{-1}	for Mg (s) + ½ O_2 (g) $\rightarrow$ MgO (s)

For a sodium chloride lattice with total radii of 212 pm:

$$U = \frac{NMZ_+Z_-}{r_0}\left[\frac{e^2}{4\pi\varepsilon_0}\right]\left(1-\frac{\rho}{r_0}\right)$$

$$U = \frac{6.022\times10^{23}\text{mol}^{-1}\times1.748\times2\times(-2)}{212\times10^{-12}\text{m}}\times2.307\times10^{-28}\,\text{Jm}\times\left(1-\frac{30}{212}\right) = -3934 \text{ kJ mol}^{-1}$$

7.20 PbS has total radii of 303 pm. For the NaCl lattice:

$$U = \frac{NMZ_+Z_-}{r_0}\left[\frac{e^2}{4\pi\varepsilon_0}\right]\left(1-\frac{\rho}{r_0}\right)$$

$$U = \frac{6.022\times10^{23}\,\text{mol}^{-1}\times1.748\times2\times(-2)}{303\times10^{-12}\,\text{m}}\times2.307\times10^{-28}\,\text{Jm}\times\left(1-\frac{30}{303}\right)$$

$$= -2888 \text{ kJ mol}^{-1}$$

$1/8\ S_8\ (s) \rightarrow S^{2-}$	535	(Energies in kJ/mol)
$Pb\ (s) \rightarrow Pb\ (g)$	196	
$Pb\ (g) \rightarrow Pb^+ + e^-$	716	
$Pb^+ \rightarrow Pb^{2+} + e^-$	1450	
$Pb^{2+} + S^{2-} \rightarrow PbS\ (s)$	U	

Total –23 kJ mol^{-1} for $Pb\ (s) + 1/8\ S_8\ (s) \rightarrow PbS\ (s)$

$U = -2{,}386$ kJ mol^{-1}, a difference of 23%.

7.21 In ZnO or TiO, additional Zn or Ti would have two more electrons than the metallic ions. As a result, any nonstoichiometry in the direction of excess Zn or Ti would supply extra electrons, making an *n*-type semiconductor.

In Cu_2S, CuI, or ZnO, excess S, I, or O would have fewer electrons than the corresponding ions. Therefore, the result of excess nonmetals in the lattice would be a *p*-type semiconductor.

7.22 Vibrational motions of the atoms in the lattice become at least partly synchronized, with positive centers moving closer together. This concentration of positive charge can attract electrons, allowing two electrons to be closer to each other than would usually be the case. When the vibrations are synchronized, this attraction can ripple through the material, helping the electrons move. Apparently the whole system acts as if it is at ground state energies, so no net change in energy is needed to keep the process going indefinitely.

7.23 The general reaction is $Na_2Z + Ca^{2+}\ (aq) \rightarrow CaZ + 2Na^+\ (aq)$, where Na_2Z is the original zeolite with sodium ions providing the positive charge. When hard water, containing Ca^{2+} or Mg^{2+}, passes through the zeolite, the ions exchange, leaving only Na^+ cations in the softened water. The zeolite can be regenerated by flushing with concentrated brine. The large Na^+ concentration reverses the reaction above.

7.24 The ion C_2^{4-} should have the following molecular orbitals (see Figure 5.7):

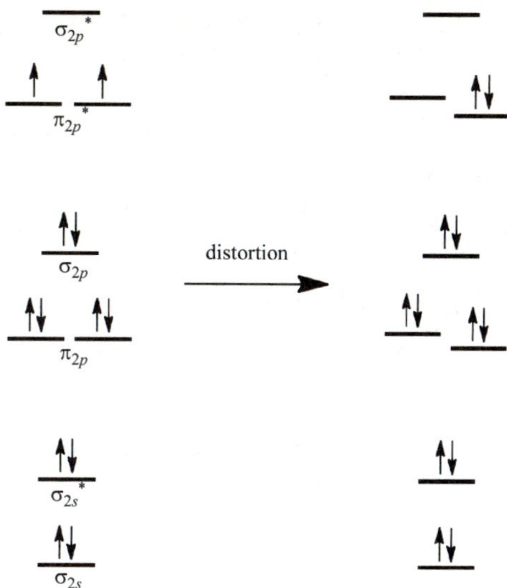

Distortion could result in removal of the degeneracy of the π_u and π_g* orbitals, giving a diamagnetic ion.

7.25 Gallium nitride has a larger band gap than gallium arsenide (continuing the trend in which gallium phosphide has a larger band gap than gallium arsenide; see Table 7.3) and emits higher energy light. Gallium nitride has grown rapidly in importance; it is used in high energy lasers, LEDs that emit light in the high energy part of the visible spectrum as well as at lower energies, and a variety of other electronics applications.

7.26 The smaller the size of the quantum dots, the greater the separation between energy levels within the dots, and the higher the energy of the photoluminescence. Consequently, the largest dots would produce the lowest energy emission bands.

7.27 First appearing in the literature in 2001, articles on medical applications of quantum dots and related topics have been more than doubling in number every two years.

7.28 **a.** $Si_4O_{12}^{8-}$

b. $Si_8O_{24}^{16-}$

c. $[Si_6O_{17}^{10-}]_n$

CHAPTER 8: CHEMISTRY OF THE MAIN GROUP ELEMENTS

8.1 **a.**

H_2	74.2 pm	436 kJ/mol
H_2^+	106 pm	255 kJ/mol

These values are consistent with the molecular orbital descriptions. H_2 has two electrons in the bonding σ orbital, H_2^+ has only one. Therefore, the attraction between the bonding electron(s) and the nuclei are weaker in the case of H_2^+, and the bond distance is longer.

b. H_3^+ was described in problems 5.20 and 5.38. It has a single pair of electrons in a 2-electron-3-center bond, with bond orders of 1/3.

8.2 He_2^+ has two electrons in the bonding σ orbital and one in the antibonding σ* orbital, with a bond order of one-half. HeH^+ has a bond order of one, but with a poorer match between the energy levels of the two atoms (–13.6 eV for H, –24.5 for He).

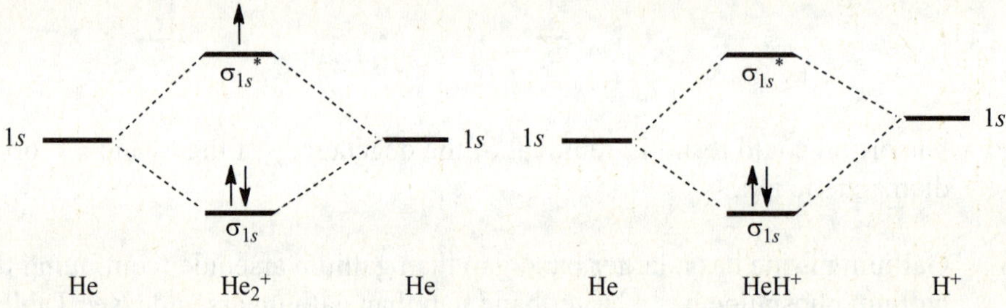

8.3 If the isoelectronic series could be continued, the ion CsF_4^+ would be expected to have a similar structure to XeF_4 and IF_4^-, with two lone pairs in *trans* positions. However, a variety of factors would make such a species truly unlikely; for example, the need for cesium to be in the chemically unknown oxidation state of 5+ and the instantaneous reaction it would undergo with the surrounding fluorines.

8.4 Figure 8.7 shows the equilibrium constants for formation of alkali metal ion complexes with cryptands, with the formation constant for the sodium complex with cryptand [2.2.1] larger than for either the lithium or potassium complexes. A similar curve is found for the alkaline earths, with the maximum at strontium. Apparently the optimum size for an ion fitting in the cryptand is larger than sodium (116 pm), but smaller than barium (152 pm) or potassium (149 pm), leaving strontium (132 pm) as the closest fit.

8.5 The diagram at the right shows the primary interactions forming molecular orbitals. The other orbitals on the fluorine atoms form lone-pair orbitals and π orbitals.

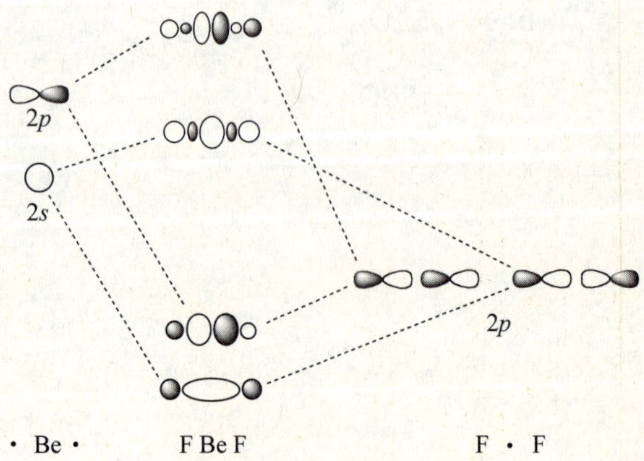

8.6 A combination of the in-plane p orbitals of Cl and the s and in-plane p orbitals of Be form the orbitals that link the Be and Cl atoms in the three-atom, two-electron bonds. Bonding interactions are shown; the corresponding antibonding orbitals are also formed, in addition to other orbitals primarily involved with the terminal chlorine atoms.

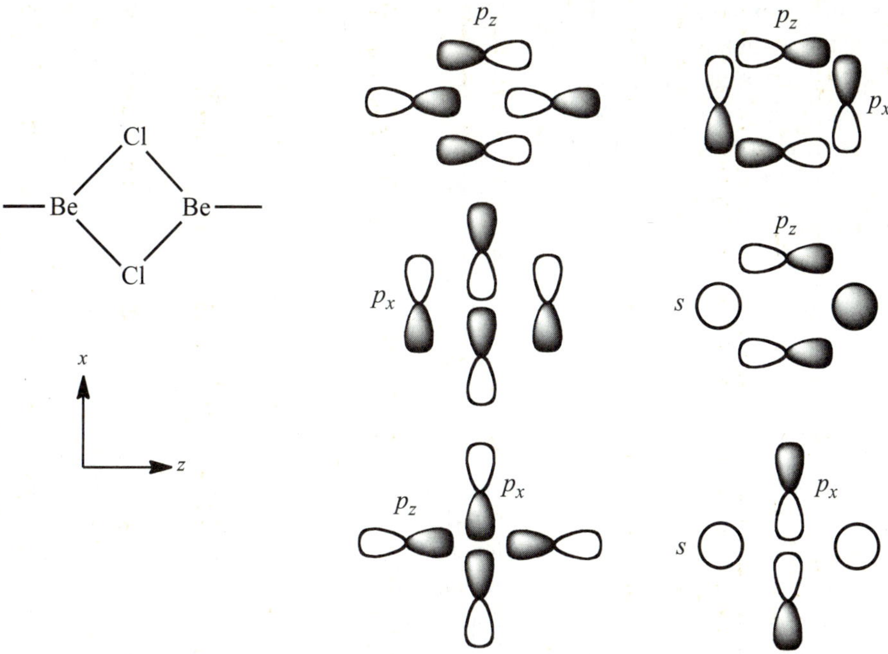

The same orbitals, shown as calculated using extended Hückel theory software, are shown below. Orbitals from the terminal chlorine atoms are included in these molecular orbitals.

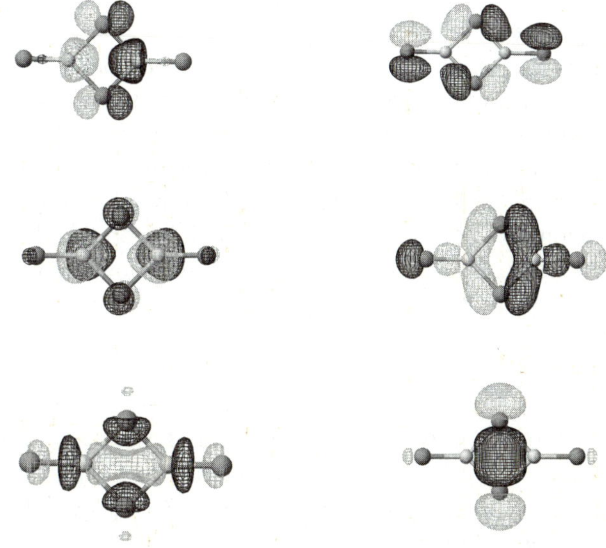

8.7 The much greater difference in orbital energies of B (–8.3 and –14.0 eV) and F (–18.7 and –40.2 eV) makes the BF bond weaker than that of CO. The *s* orbital of B and the *p* orbital of F can interact, but the *p–p* combination cannot interact significantly because of the large difference in *p* orbital energies. This mismatch results in weak single bonding, rather than the triple bond of CO.

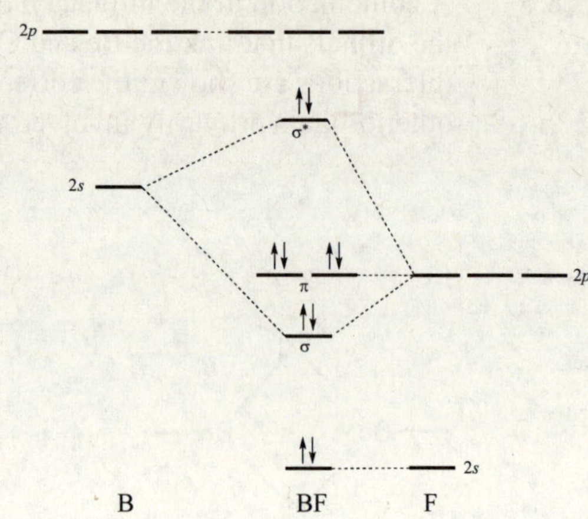

8.8 **a.** Evidence cited for a double bond includes an X-ray crystal structure which shows a very short boron–boron distance (156 pm) that is considerably shorter than compounds having boron–boron single bonds and comparable to the bond distance in compounds with double bonds between borons. Density functional theory computations also showed a HOMO with boron–boron π bonding.

R = isopropyl

b. The product in this case was a silicon compound having a structure similar to the boron compound shown above, but lacking hydrogens; nevertheless, it has bent geometry at the silicons, leading to the zigzag structure shown at right. An X-ray crystal structure similarly shows a silicon–silicon distance consistent with a double bond, and density functional theory yields a HOMO with π bonding between the silicon atoms.

8.9 The combination of orbitals forming the bridging orbitals in $Al_2(CH_3)_6$ is similar to those of diborane (pp. 272–273). The difference is that the CH_3 group has a *p* orbital (or sp^3 hybrid orbital) available for bonding to the aluminum atoms, with the same symmetry possibilities as those of H atoms in B_2H_6.

8.10

D_{2h}	E	$C_2(z)$	$C_2(y)$	$C_2(x)$	i	$\sigma(xy)$	$\sigma(xz)$	$\sigma(yz)$	
$\Gamma(p_z)$	2	2	0	0	0	0	2	2	
$\Gamma(p_x)$	2	–2	0	0	0	0	2	–2	
$\Gamma(1s)$	2	0	0	2	0	2	2	0	
A_g	1	1	1	1	1	1	1	1	z^2
B_{2g}	1	–1	1	–1	1	–1	1	–1	xz
B_{1u}	1	1	–1	–1	–1	–1	1	1	xy
B_{3u}	1	–1	–1	1	–1	1	1	–1	x

Reduction of the representations Γ gives the following:

Boron orbitals:

a. $\Gamma(p_z) = A_g + B_{1u}$

b. $\Gamma(p_x) = B_{2g} + B_{3u}$

Hydrogen orbitals:

c. $\Gamma(1s) = A_g + B_{3u}$

d. Treating each group orbital as a single orbital, the orbitals below have the indicated symmetries.

D_{2h}	E	$C_2(z)$	$C_2(z)$	$C_2(z)$	i	$\sigma(xy)$	$\sigma(xz)$	$\sigma(yz)$
$A_g\ (p_z)$	1	1	1	1	1	1	1	1
$B_{1u}\ (p_z)$	1	1	–1	–1	–1	–1	1	1
$B_{2g}(p_x)$	1	–1	1	–1	1	–1	1	–1
$B_{3u}\ (p_x)$	1	–1	–1	1	–1	1	1	–1
$A_g\ (s)$	1	1	1	1	1	1	1	1
B_{3u}	1	–1	–1	1	–1	1	1	–1

8.11 Carbon has two lone pairs in $C(PPh_3)_2$, with VSEPR predicting a tetrahedral angle. The bulky phenyl groups force a larger angle.

8.12 Comparing the compounds CaC_2, CeC_2, and YC_2, the unusual 2+ group 3 ions each have an extra electron. It has been suggested (Greenwood and Earnshaw, *Chemistry of the Elements*, 2nd ed., p. 299) that there is some transfer of this extra electron to the π* orbitals of the dicarbide (or acetylide) ion, resulting in a longer bond.

8.13 Radioactive decay obeys a first order kinetic equation:

$$\frac{dx}{dy} = -kx; \quad \ln\left(\frac{x}{x_0}\right) = -kt$$

The relationship of the half-life (the time at which $x = \frac{1}{2} x_o$) and the rate constant is

$$\ln\left(\frac{x}{x_0}\right) = \ln\left(\frac{1}{2}\right) = -kt_{1/2} = -0.693$$

$$k = \frac{0.693}{t_{1/2}}$$

$$k = \frac{0.693}{t_{1/2}} = 1.21 \times 10^{-4}\, y^{-1}$$

$$\ln\left(\frac{x}{x_0}\right) = \ln(.56) = -1.21 \times 10^{-4}\, t$$

$$t = 4.8 \times 10^3\, y$$

8.14 I_h symmetry includes 12 C_5 axes, 12 C_5^2 axes, 20 C_3 axes, 15 C_2 axes, an inversion center, 12 S_{10} axes, 20 S_6 axes, and 15 mirror planes. One of the C_5 axes and an S_{10} axis can be seen in the middle of the second fullerene figure (end view) in Figure 8.16. There are six of each of these, each representing two rotation axes, C_5 and C_5^4, for a total of 12. The same axes fit the 12 S_5^2 axes. If the structure is rotated to line up two hexagons surrounded by alternating pentagons and hexagons, the 20 C_3 and 20 S_6 axes can be seen (one in the center, three in the first group around that hexagon, and three adjacent pairs in the next group, doubled because of C_3 and C_3^2). The C_2 axes pass through bonds shared by two hexagons, with pentagons at each end. These original hexagons each have two more, making 5, the next ring out has 8, and the perimeter (seen edge on) has two for a total of 20 C_2. There are five mirror planes through the center of the pentagons surrounded by five hexagons (see the end view of Figure 8-16 again), and there are three sets of these for a total of 15. (All this is best seen using a model!)

8.15

a. T_d

b. I_h

c. D_{5h}

d. zigzag: C_{2h} armchair: D_{2h}

8.16 Obtained from the hydrogenation of graphene, graphane was apparently first reported in early 2009 (see J. Agbenyega, *Mater. Today*, **2009**, *12*, 13), and research interest has grown rapidly.

8.17 Printing on transparencies is suggested for this exercise. Rolling up of the diagram should show that more than one chiral form is possible.

8.18 Readers are encouraged to search for the most recent references in this promising area. At this writing (September 2010) the reference has been cited in at least 42 other articles in scientific journals.

8.19 The increased stability of 2+ oxidation states as compared to 4+ is an example of the "inert pair" effect (see Greenwood and Earnshaw, *Chemistry of the Elements*, 2nd ed., pp. 226, 227, 374). In general, the ionization energy decreases going down a column of the periodic table, because of greater shielding by the inner electrons. In this family, removal of the second electron is fairly easy, as it is the first in the higher energy *p* orbitals. However, the next two electrons to be removed are the *s* electrons, and they are not as thoroughly shielded in the ions. The effect is larger for the three lower members of the group because the d^{10} electrons are added between the *s* and *p* electrons. The lower electronegativity of C and Si make them more likely to form covalent bonds than ions.

8.20 **a.** $H_2ISiSiH_2I$ has a C_2 axis perpendicular to the plane of the diagram as drawn, a perpendicular mirror plane (the plane of the diagram), and an inversion center. C_{2h}.

b. The Si–H stretches have the reducible representation shown below, which reduces to $\Gamma = A_g + B_g + A_u + B_u$. A_u and B_u are IR-active.

C_{2h}	E	C_2	i	σ_h	
Γ	4	0	0	0	
A_g	1	1	1	1	R_z
B_g	1	–1	1	–1	R_x, R_y
A_u	1	1	–1	–1	z
B_u	1	–1	–1	1	x, y

8.21 $P_4\,(g) \rightleftharpoons 2\,P_2\,(g)$, $\Delta H = 217$ kJ mol^{-1}
P_4 has six P–P bonds, so six bonds are broken and two triple bonds are formed.

ΔH = Σbond dissociation energy (reactant) – Σbond dissociation energy (product)
$217 = 6 \times 200 - 2$ (bond dissociation energy of P≡P)

Bond dissociation energy of P≡P = 492 kJ mol^{-1}

The p_π orbitals in P_2 do not overlap as effectively as those in N_2, resulting in weaker π bonds in P_2.

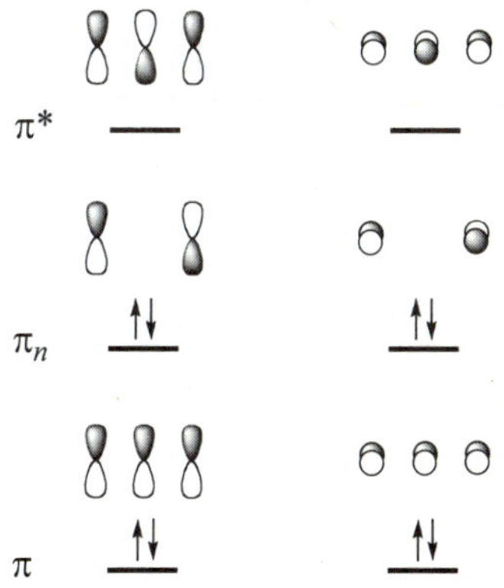

8.22 **a.** N_3^- has molecular orbitals similar to those of CO_2 (pp. 152–157), with two occupied σ orbitals and two occupied π orbitals for a total of 4 bonds. Here the atomic orbitals have identical energies.

π^*

π_n

π

b. Because the HOMOs of N_3^- are primarily composed of *p* orbitals of the terminal nitrogen atoms, H^+ bonds at an angle to the N=N=N axis. The angle is 114°, larger than the simple 90° predicted by bonding to a *p* orbital.

c. The H–N–N angle is nearly that of sp^2 hybrids. Such bonding reduces the importance of the first resonance structure shown and increases the importance of the second, with the triple bond between the end nitrogens. In MO terms, the proton draws electron density away from the nearby N–N bond, weakening it and strengthening the bond at the opposite end of the molecule.

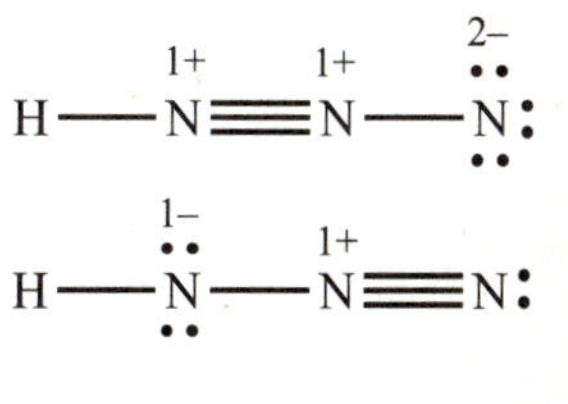

$\mathrm{H{-}\overset{1+}{N}{\equiv}\overset{1+}{N}{-}\overset{2-}{\ddot{N}}\!:}$

$\mathrm{H{-}\overset{1-}{\ddot{N}}{-}\overset{1+}{N}{\equiv}N\!:}$

8.23 Hydrazine has two positive hydrogens on each nitrogen, ammonia has three. This leaves the ammonia nitrogen more negative, and therefore more basic.

8.24 The larger central atoms force the bonded pairs farther away from the central nucleus, reducing their repulsive force. The lone pair remains closer to the nucleus, and therefore has more influence and forces a smaller angle for the molecules with larger central atoms.

8.25 One *p* orbital of each oxygen is used for the σ bond to nitrogen, and the *p* orbital perpendicular to the plane of the molecule is used for a π orbital. The remaining *p* orbital is in the plane of the molecule, so addition of a proton to this lone pair leaves the entire molecule planar.

8.26 From Table 8.10

N_2O	$C_{\infty v}$	NO	$C_{\infty v}$	NO_2	C_{2v}	N_2O_3	C_s
N_2O_4	D_{2h}	N_2O_5	D_{2h}	NO^+	$C_{\infty v}$	NO_2^+	$D_{\infty h}$
NO_2^-	C_{2v}	NO_3^-	D_{3h}	$N_2O_2^{2-}$	C_{2h}	NO_4^{3-}	T_d
HNO_2	C_s	HNO_3	C_s				

8.27 The bonding between O_2 units is proposed to involve interactions between the singly occupied π* orbitals of neighboring molecules. Such interactions would generate both bonding and antibonding molecular orbitals, with the originally unpaired π* electrons stabilized in energy as they occupy the bonding orbitals in O_8. One such interaction is shown at right. Diagrams of the four highest occupied molecular orbitals of O_8 (which may be viewed as $(O_2)_4$) are shown in the reference.

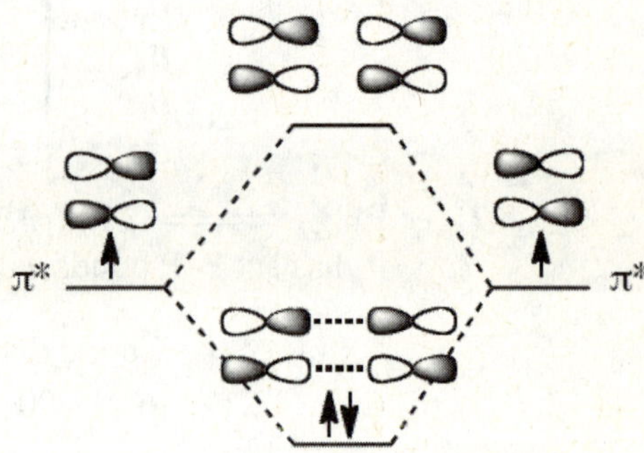

8.28 S_2 is similar to O_2, with a double bond. As a result, the bond is shorter than single bonds in S_8.

8.29 MnF_6^{2-} acts as a Lewis base, donating two F^- ions to SbF_5 molecules:

$$MnF_6^{2-} + 2\ SbF_5 \rightarrow 2\ SbF_6^- + MnF_4$$

The remaining MnF_4 can then lose an additional F atom, with two combining to form F_2:

$$2\ MnF_4 \rightarrow F_2 + 2\ MnF_3$$

8.30 **a.** Although the ClO_3 groups in Cl_2O_7 are highly electronegative, their size is apparently responsible for the larger angle in Cl_2O_7 (118.6°) than in Cl_2O (110.9°).

O: Cl–O–Cl 110.9° O_3Cl–O–ClO_3 118.6° $[O_3Cr$–O–$CrO_3]^{2-}$ 126°

b. In the dichromate ion the highly electropositive Cr atoms allow the central oxygen to attract electrons more strongly than is possible in Cl_2O_7, resulting in stronger electron–electron repulsions around the central oxygen in $[Cr_2O_7]^{2-}$ and a larger bond angle (126°).

8.31 I_3^- is linear because there are three lone pairs in a trigonal geometry on the central I. I_3^+ is bent because there are only two lone pairs on the central I.

8.32 B has only three valence electrons. Adding six from the hydrogen atoms gives B_2H_6 a total of 12 electrons. The 3-center 2-electron bonds results in four pairs around each boron atom and nearly tetrahedral symmetry at each boron atom. Iodine has seven valence electrons initially. In I_2Cl_6, five more are added to each iodine, resulting in 12 electrons and octahedral geometry around each iodine.

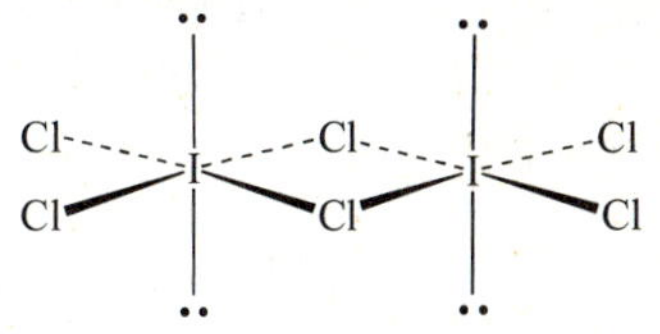

8.33 $F^- + BrF_3 \rightarrow BrF_4^-$ KF acts as a base; BrF_4^- is the solvent anion.

$SbF_5 + BrF_3 \rightarrow BrF_2^+ + SbF_6^-$ BrF_2^+ is the cation of the solvent; SbF_5 acts as the acid.

8.34 **a.** Br_2^+ and I_2^+ have one less antibonding electron than Br_2 and I_2, so the cations have bond orders of 1.5 and shorter bonds than the neutral molecules:

	Bond Order	Bond Distance (pm)		Bond Order	Bond distance (pm)
Br_2	1.0	228	I_2	1.0	267
Br_2^+	1.5	213	I_2^+	1.5	256

b. The most likely transition is from the π_g^* (HOMO) to σ_u^* (LUMO) (see Figure 5.7 for comparable energy levels of F_2). Because the observed colors are complementary to the colors absorbed, Br_2^+ is expected to absorb primarily green light, and I_2^+ is expected to absorb primarily orange light. Because orange light is less energetic than green, I_2^+ has the more closely spaced HOMO and LUMO.

c. In both I_2 and I_2^+ the HOMO and LUMO are the π_g^* and σ_u^*, respectively. I_2 absorbs primarily yellow light (to give the observed violet color) and I_2^+ absorbs primarily orange (to give blue). Because yellow light is more energetic than orange, I_2 has the more widely spaced HOMO and LUMO.

8.35 **a.** In I_2^+ there are three electrons in the π^* orbitals that result from interactions between 5*p* orbitals (these are the orbitals labeled $4\pi_g^*$ in Figure 6.9). Interaction between the singly occupied π^* orbitals of two I_2^+ ions can lead to stabilization of electrons by formation of a bonding molecular orbital that can connect the I_2^+ units when I_4^{2+} is formed.

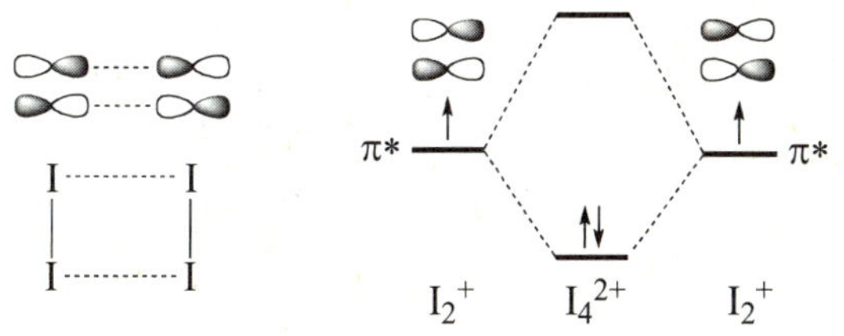

b. Higher temperature would provide the energy to break the bonds between the I_2^+ units and favor the monomer. In addition, formation of two I_2^+ ions from I_4^{2+} would be accompanied by an increase in entropy, also favoring the monomer at high temperature.

8.36 There are three possibilities:

The third structure, with the lone pair and double bonds in a facial arrangement, is least likely because it would have the greatest degree of electron–electron repulsions involving these regions of high electron concentrations.

C_{2v} C_s C_s

Infrared data can help distinguish between the other two structures. In the first structure, only the antisymmetric stretch would be IR-active, giving a single absorption band. In the second structure, both the symmetric and antisymmetric stretches would be IR-active, giving two bands. Observation of two I–O bands is therefore consistent with the second structure. This structure, which has fewer 90° lone pair–double bond repulsions than the first structure, is also expected to be the most likely by VSEPR considerations. Other experimental data are also consistent with the second structure.

8.37 These reactions take place in the gas phase. The initial product of Xe with PtF_6 is believed to be $Xe^+ PtF_6^-$; however, when these two ions are in close proximity, they may react further to give $[XeF]^+$, $[Pt_2F_{11}]^-$, and other products. SF_6, if present in large excess, prevents the formation of these secondary products, apparently by acting as an inert diluent and preventing effective collisions between the desired products.

8.38

C_{2v} C_{4v} C_{2v} D_{3h}

8.39 Two electrons are in the bonding orbital and two are in the nonbonding orbital:

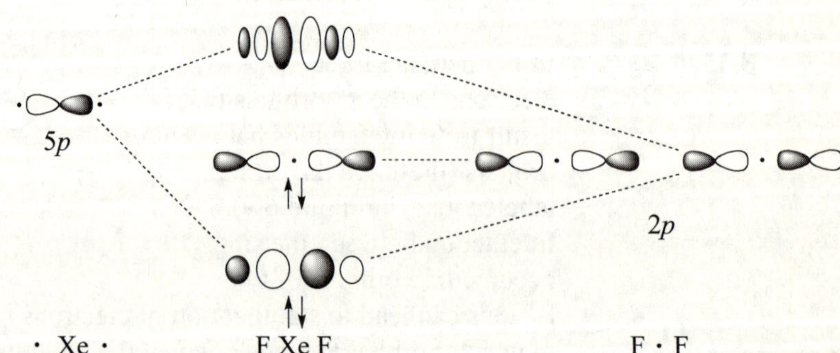

8.40 $Xe(OTeF_5)_4$: The $OTeF_5$ group has one electron available for bonding on the O, so the four groups form a square planar structure around Xe, with lone pairs in the axial positions. $O{=}Xe(OTeF_5)_4$: A square pyramidal structure, with O in one of the axial positions and a lone pair on the other, and $OTeF_5$ groups in the square base.

8.41 Half reactions:

$$Mn^{2+} + 4H_2O \rightarrow MnO_4^- + 8H^+ + 5e^-$$
$$XeO_6^{4-} + 12H^+ + 8e^- \rightarrow Xe + 6H_2O$$

Overall reaction:

$$8Mn^{2+} + 5XeO_6^{4-} + 2H_2O \rightarrow 8MnO_4^- + 4H^+ + 5Xe$$

8.42 XeF_5^- has D_{5h} symmetry. The reducible representation for Xe–F stretching is

D_{5h}	E	$2C_5$	$2C_5^2$	$5C_2$	σ_h	$2S_5$	$2S_5^3$	$5\sigma_v$
Γ	5	0	0	1	5	0	0	1

which reduces to $\Gamma = A_1' + E_1' + E_2'$, with only E_1' IR-active.

8.43 **a.** Point group: C_{4v}

C_{4v}	E	$2C_4$	C_2	$2\sigma_v$	$2\sigma_d$	
Γ	18	2	–2	4	2	
A_1	1	1	1	1	1	z
A_2	1	1	1	–1	–1	R_z
B_1	1	–1	1	1	–1	
B_2	1	–1	1	–1	1	
E	2	0	–2	0	0	$(x, y), (R_x, R_y)$

b. $\Gamma = 4A_1 + A_2 + 2B_1 + B_2 + 5E$

c. Translation: $A_1 + E$ (match x, y, and z)
Rotation: $A_2 + E$ (match R_x, R_y, and R_z)
Vibration: all that remain: $3A_1 + 2B_1 + B_2 + 3E$

8.44 By the VSEPR approach, XeF_2^{2-} would be expected to have a steric number of 6, with four lone pairs on xenon. Two structures, *cis* and *trans*, need to be considered. In the *cis* structure there would be five lone pair–lone pair interactions at 90°, and in the *trans* structure there would be only four such interactions. The *trans* structure, therefore, would be expected to be more likely. However, the number of lone pair–lone pair repulsions would still be high, likely making this a difficult ion to prepare.

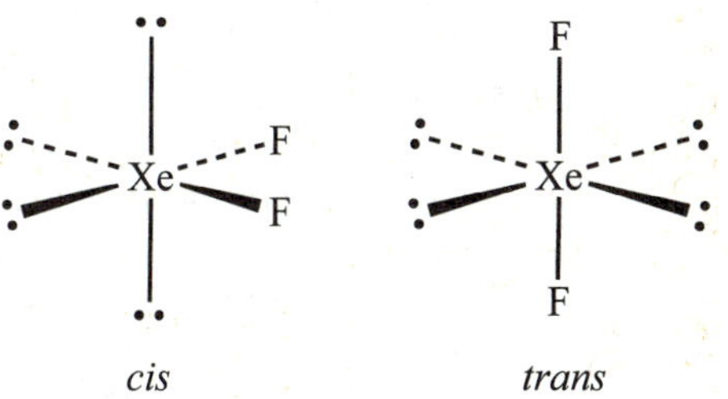

8.45 **a.** The $[XeF]^+$ ion, with a total of 14 valence *s* and *p* electrons, would be expected to have a bond order of 1 (see Section 5.2.1).

b. Using the VSEPR approach, the following structures can be drawn for the cations of compounds **1-4**:

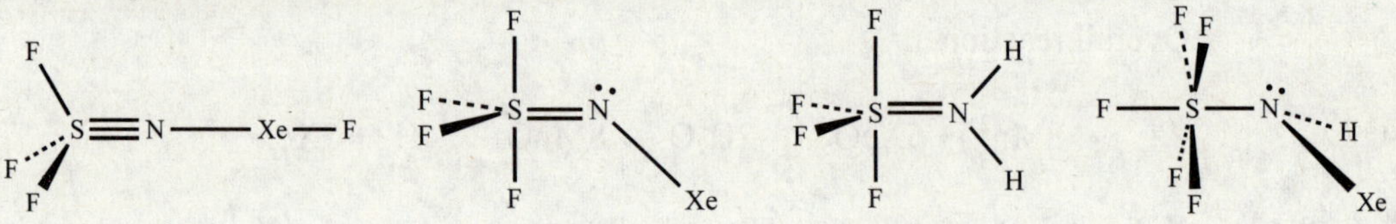

The $[F_3NXeF]^+$ ion, **1**, with a triple bond between sulfur and nitrogen, has the shortest N–S distance (139.7 pm), and $[F_5SN(H)Xe]^+$, with a single bond, has the longest distance (176.1 pm). The other ions have bond distances between these extremes that are consistent with double bonds between sulfur and nitrogen.

c. Compound **1** is expected to have a cation with linear bonding around nitrogen, as shown. However, its crystal structure shows significant bending, with an S–N–Xe angle of 142.6°. This bending is attributed to close N···F contacts within the crystalline lattice.

d. VSEPR would predict linear bonding with three lone pairs on xenon as, for example, in XeF_2. The angle measured by X-ray crystallography is 179.6°.

e. With significant S–N double bonding, these groups are equatorial.

f. The S–F_{axial} bonds are longer, as in many other VSEPR examples (see Figures 3.17 through 3.19). Average distances are:

Ion	S–F (axial)	S–F (equatorial)
2	158.2 pm	152.4 pm
3	156.1 pm	151.8 pm

8.46 $[FBeNe]^+$ The relative order of molecular orbitals may vary depending on the software used. A simple approach, using an extended Hückel calculation, gave the following results:

F–Be bonding

Several orbitals involved:

HOMO: degenerate pair — π bonding between Be and F

Next orbital below HOMO — σ bonding by the F p_z and the Be *s*

Be–Ne bonding

One molecular orbital contributing significantly:

Four orbitals below HOMO — σ bonding by the Ne p_z and the Be *s* and p_z

Overall, the interactions between Be and F are stronger than between Be and Ne.

8.47 In Xe_2^+, there are a total of 54 orbitals. We will classify only the higher-energy orbitals here, starting with the highest energy (as calculated using the relatively simple extended Hückel approach). The orbitals used are the 5*s*, 5*p*, and 4*d*.

Molecular Orbital	Type of Interaction	Atomic Orbitals	Molecular Orbital	Type of Interaction	Atomic Orbitals
HOMO	σ^*	p_z	HOMO–8 HOMO–9	δ^*	$d_{x^2-y^2}$, d_{xy}
HOMO –1 HOMO –2	π^*	p_x p_y	HOMO–10 HOMO–11	δ	$d_{x^2-y^2}$, d_{xy}
HOMO –3	σ	p_z	HOMO–12 HOMO–13	π^*	d_{xz}, d_{yz}
HOMO–4 HOMO–5	π	p_x p_y	HOMO–14 HOMO–15	π	d_{xz}, d_{yz}
HOMO–6	σ^*	s	HOMO–16	σ^*	d_{z^2}
HOMO–7	σ	s	HOMO–17	σ	d_{z^2}

If this were the neutral Xe_2, there would be no bond, because every occupied bonding orbital would be offset by an occupied antibonding orbital. In this case, one electron is missing, so there is ½ bond, making for a long bond. In addition, Xe is a large atom, so the bond would naturally be long.

Overlap for the δ and δ^* orbitals is very small in this case because there are filled 5*s* and 5*p* orbitals outside the 4*d* orbitals.

8.48 The reference by Steudel and Wong provides illustrations for the four highest occupied molecular orbitals. In addition to the interesting symmetry of these orbitals, the way in which the π^* orbitals of O_2 are imbedded in the molecular orbitals of O_8 should be noted.

CHAPTER 9: COORDINATION CHEMISTRY I: STRUCTURES AND ISOMERS

9.1

Hexagonal:	C_{2v}	C_{2v}	D_{2h}
Hexagonal pyramidal:	C_s	C_s	C_{2v}
Trigonal prismatic:	C_s	C_{2v}	C_2
Trigonal antiprismatic:	C_s	C_2	C_{2h}

The structures with C_2 symmetry would be optically active.

9.2 **a.** dicyanotetra(methylisocyano)iron(0) or dicyanotetra(methylisocyano)iron(0) Both methods of naming result in the same name in this case.

b. rubidium tetrafluoroargentate(III) or rubidium tetrafluoroargentate(1–)

c. *cis*- and *trans*-carbonylchlorobis(triphenylphosphine)iridium(I) or *cis*- and *trans*-carbonylchlorobis(triphenylphosphine)iridium(0)

d. pentaammineazidocobalt(III) sulfate or pentaammineazidocobalt(2+) sulfate

e. diamminesilver(I) tetrafluoroborate(III) or diamminesilver(1+) tetrafluoroborate(1–) (The BF_4^- ion is commonly called simply "tetrafluoroborate.")

9.3 **a.** tris(oxalato)vanadate(III) or tris(oxalato)vanadate(3–) (In the first printing, problem 9.2.d. was duplicated here.)

b. sodium tetrachloroaluminate(III) or sodium tetrachloroaluminate(1–)

c. carbonatobis(ethylenediamine)cobalt(III) chloride or carbonatobis(ethylenediamine)cobalt(1+) chloride

d. tris(2,2′-bipyridine)nickel(II) nitrate or tris(2,2′-bipyridine)nickel(2+) nitrate (The IUPAC name of the bidentate ligand, 2,2′-bipyridyl may also be used; this ligand is most familiarly called "bipy.")

e. hexacarbonylmolybdenum(0) (also commonly called "molybdenum hexacarbonyl") The (0) is often omitted. (In the first printing, problem 9.3.c. was duplicated here.)

9.4 **a.** tetraamminecopper(II) or tetraamminecopper(2+)

b. tetrachloroplatinate(II) or tetrachloroplatinate(2–)

c. tris(dimethyldithiocarbamato)iron(III) or tris(dimethyldithiocarbamato)iron(0)

d. hexacyanomanganate(II) or hexacyanomanganate(4–)

e. nonahydridorhenate(VII) or nonahydridorhenate(2–) (This ion is commonly called "enneahydridorhenate.")

9.5 **a.** triamminetrichloroplatinum(IV) or triamminetrichloroplatinum(1+)

b. diamminediaquadichlorocobalt(III) or diamminediaquadichlorocobalt(1+)

c. diamminediaquabromochlorocobalt(III) or diamminediaquabromochlorocobalt(1+)

d. triaquabromochloroiodochromium(III) or triaquabromochloroiodochromium(0)

e. dichlorobis(ethylenediamine)platinum(IV) or dichlorobis(ethylenediamine)platinum(2+) or dichlorobis(1,2-ethanediamine)platinum(IV) or dichlorobis(1,2-ethanediamine)platinum(2+)

f. diamminedichloro(*o*-phenanthroline)chromium(III) or diamminedichloro(*o*-phenanthroline)chromium(1+) or diamminedichloro(1,10-phenanthrolene)chromium(III) or diamminedichloro(1,10-phenanthrolene)chromium(1+)

g. bis(bipyridine)bromochloroplatinum(IV) or bis(bypyridine)bromochloroplatinum(2+) or bis(2,2′-bipyridyl)bromochloroplatinum(IV) or bis(2,2′-bipyridyl)bromochloroplatinum(2+)

h. dibromo[*o*-phenylene(dimethylarsine)(dimethylphosphine)]rhenium(II) or dibromo[*o*-phenylene(dimethylarsine)(dimethylphosphine)]rhenium(0) or dibromo[1,2-phenylene(dimethylarsine)(dimethylphosphine)]rhenium(II) or dibromo[1,2-phenylene(dimethylarsine)(dimethylphosphine)]rhenium(0)

i. dibromochlorodiethylenetriaminerhenium(III) or dibromochlorodiethylenetriaminerhenium(0) or dibromochloro(2,2′-diaminodiethylamine)rhenium(III) or dibromochloro(2,2′-diaminodiethylamine)rhenium(0)

9.6 **a.** *cis*-dicarbonylbis(dimethyldithiocarbamato)ruthenium(III) or *cis*-dicarbonylbis(dimethyldithiocarbamato)ruthenium(0)

b. trisoxalatocobaltate(III) or trisoxalatocobaltate(3–)

c. tris(ethylenediamine)ruthenium(II) or tris(ethylenediamine)ruthenium(2+)

d. bis(2,2′-bipyridine)dichloronickel(II) or bis(2,2′-bipyridine)dichloronickel(2+)

9.7 **a.** Bis(en)Co(III)-μ-amido-μ-hydroxobis(en)Co(III)

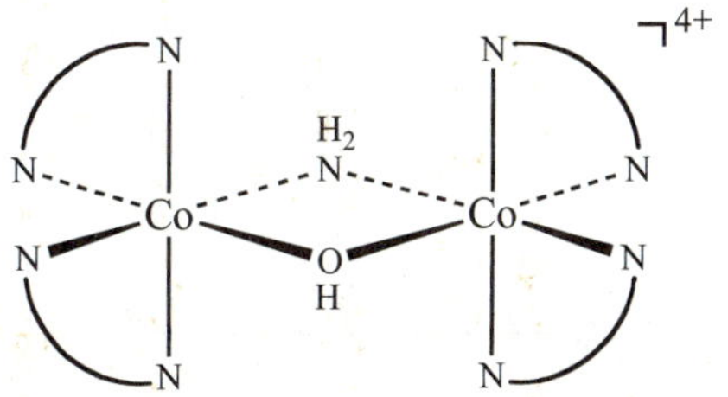

b. Diaquadiiododinitritopalladium(IV)

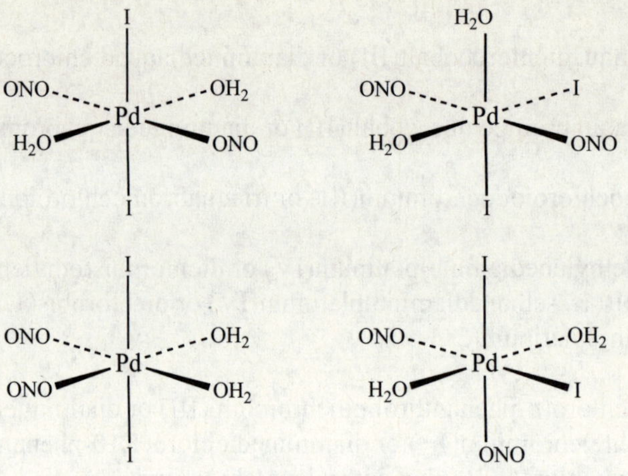

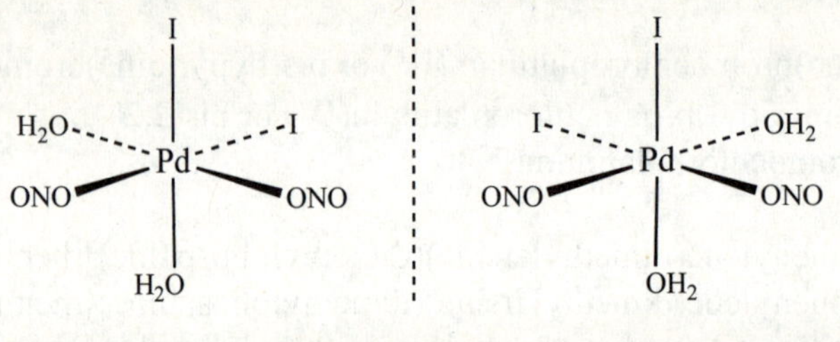

I, H2O, ONO, OH2, Pd, ONO, I

enantiomers

c. $Fe(dtc)_3$

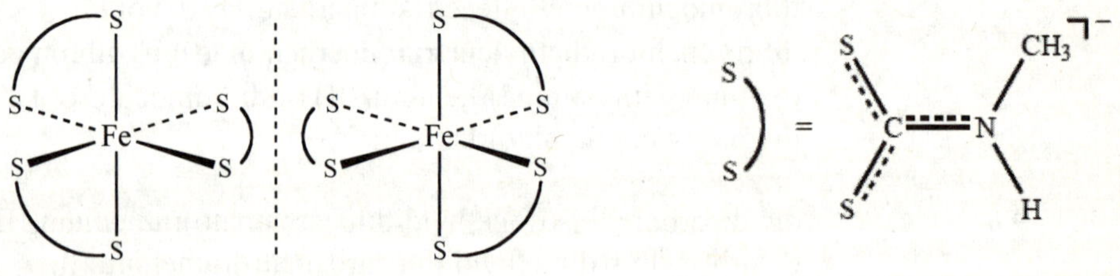

At low temperature, restricted rotation about the C–N bond can lead to additional isomers as a consequence of the different substituents on the nitrogen. These isomers, which are of the type in problem 9.11, can be observed by NMR.

9.8 **a.** triammineaquadichlorocobalt(III) chloride   Isomers are of the cation:

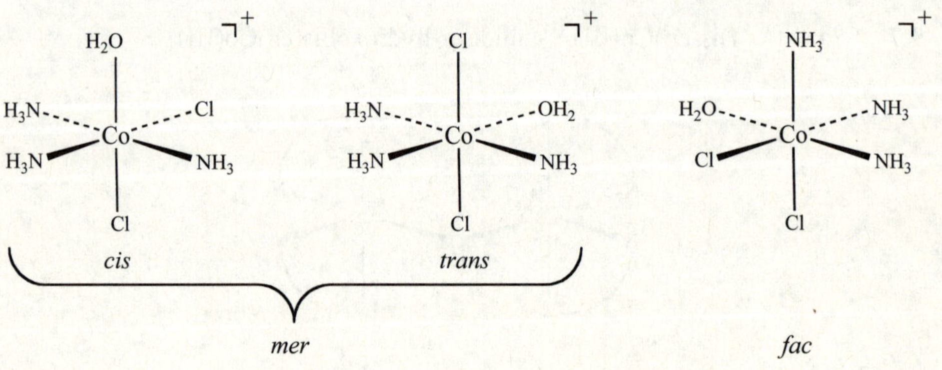

cis *trans*

mer

fac

b. μ-oxo-bis(pentammine-chromium(III)) ion

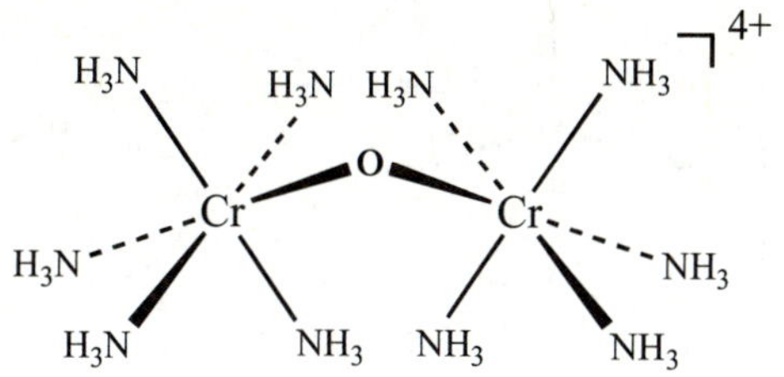

H_3N H_3N H_3N NH_3 $4+$ O Cr Cr H_3N NH_3 H_3N NH_3 NH_3 NH_3

c. potassium diaquabis(oxalato)manganate (III) Isomers are of the anion:

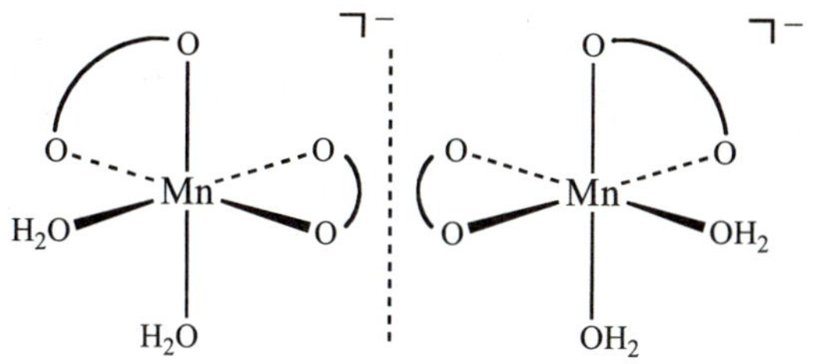

O Mn O O H_2O O H_2O | O Mn O O O OH_2 OH_2 | H_2O O Mn O O O H_2O

cis enantiomers 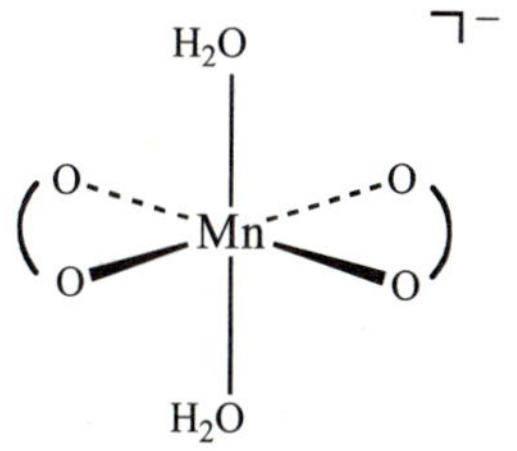 *trans*

9.9 **a.** *cis*-diamminebromochloroplatinum(II)

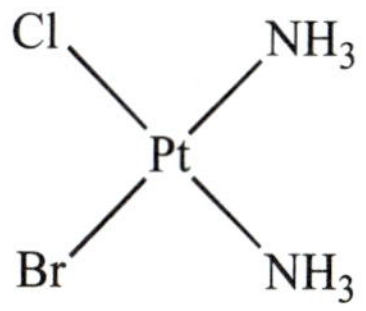

Cl NH_3 Pt Br NH_3

b. diaquadiiododinitritopalladium(IV)

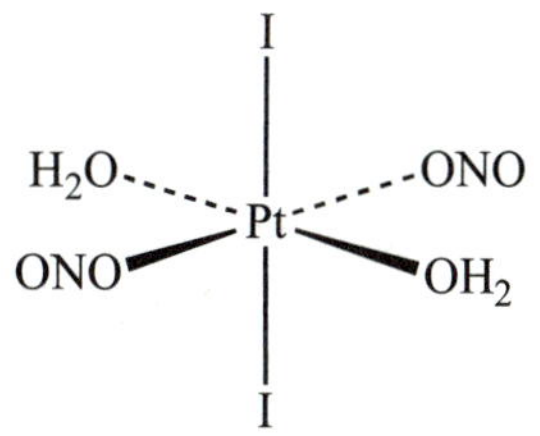

I H_2O ONO Pt ONO OH_2 I

c. tri-μ-carbonylbis(tricarbonyliron(0))

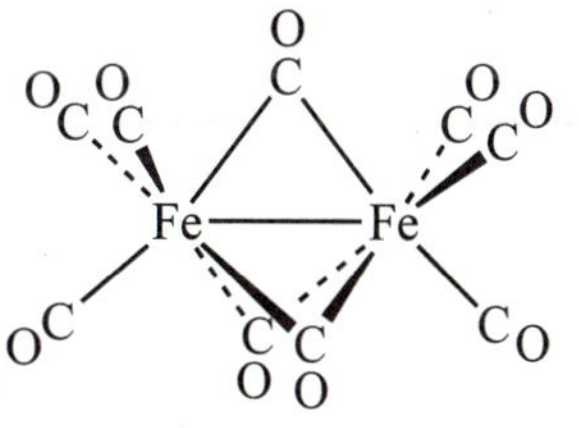

OC CO CO Fe Fe CO OC CO CO CO CO

9.10

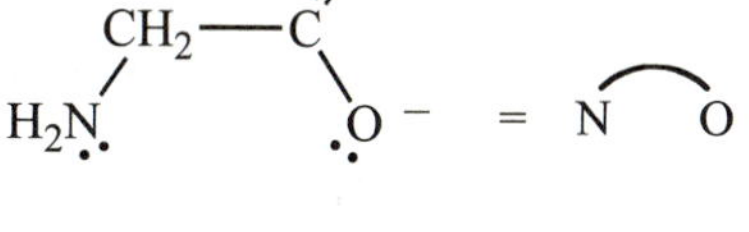

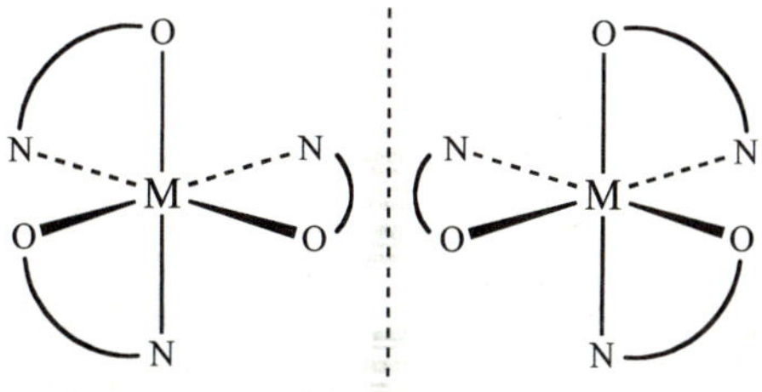

CH_2—C(=O) H_2N O^- = N⌒O

N O M N O O N | N O M N O O N | O N M N O N O | N O M N O N O

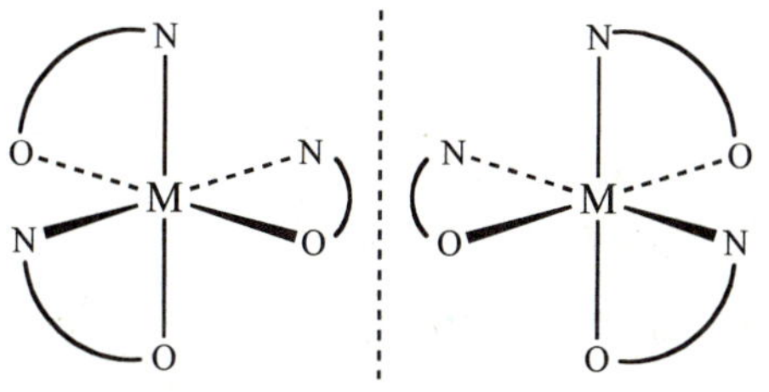

9.11 $M(AB)_3$ *fac* *mer*

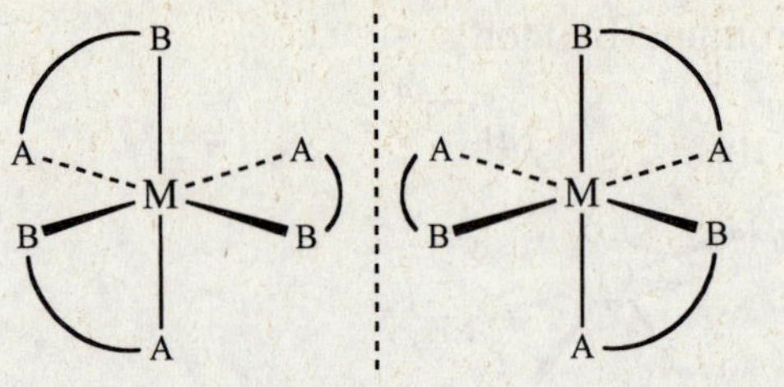

fac 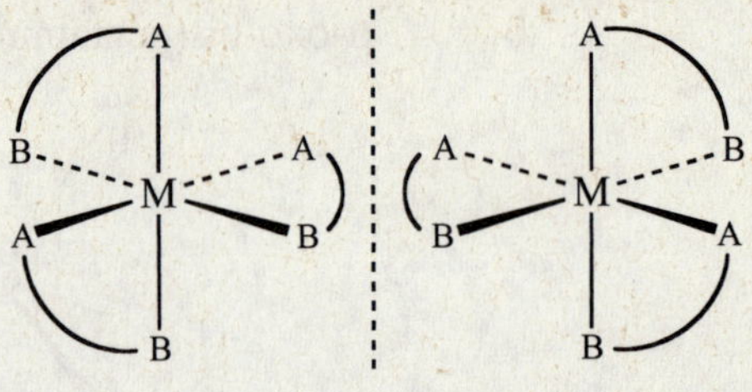 mer

9.12 a. $[Pt(NH_3)_3Cl_3]^+$

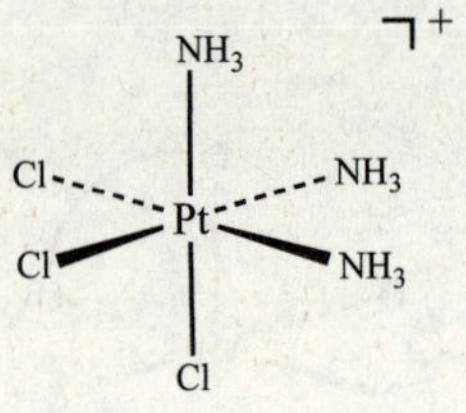

fac 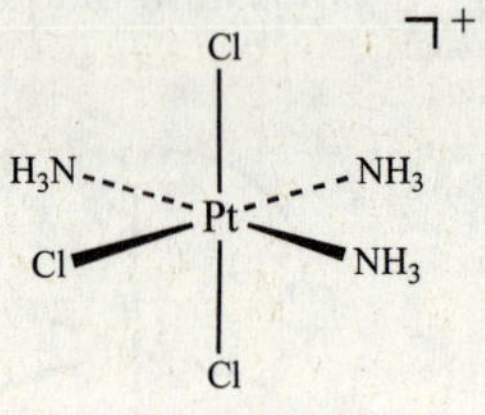 mer

b. $[Co(NH_3)_2(H_2O)_2Cl_2]^+$

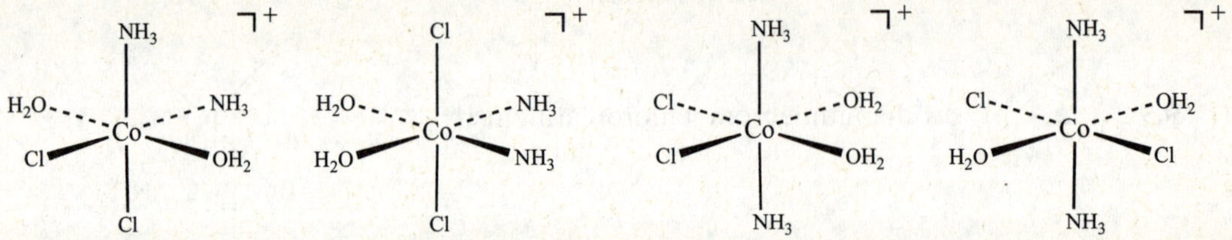

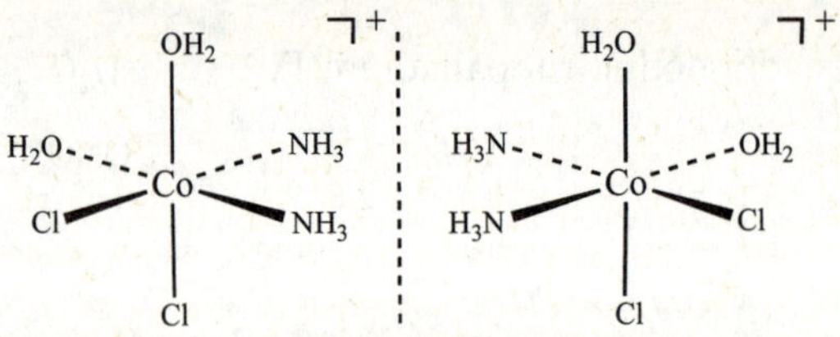

enantiomers

c. $[Co(NH_3)_2(H_2O)_2BrCl]^+$

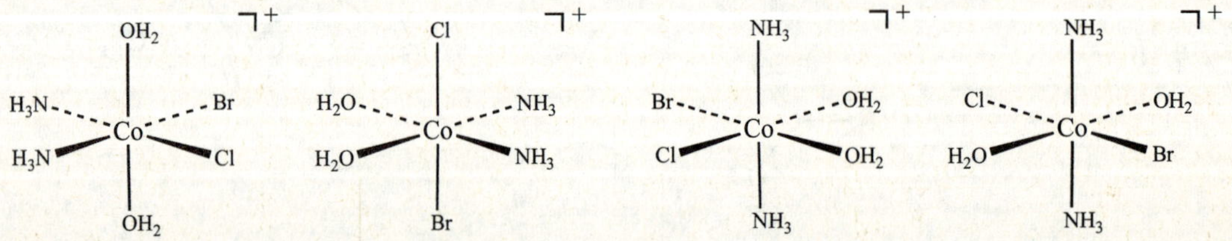

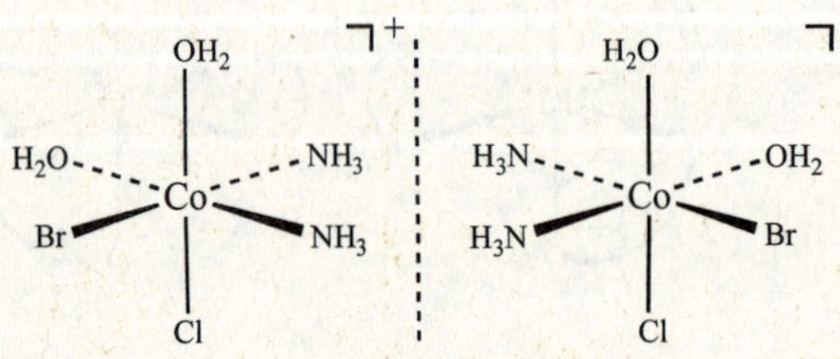

enantiomers 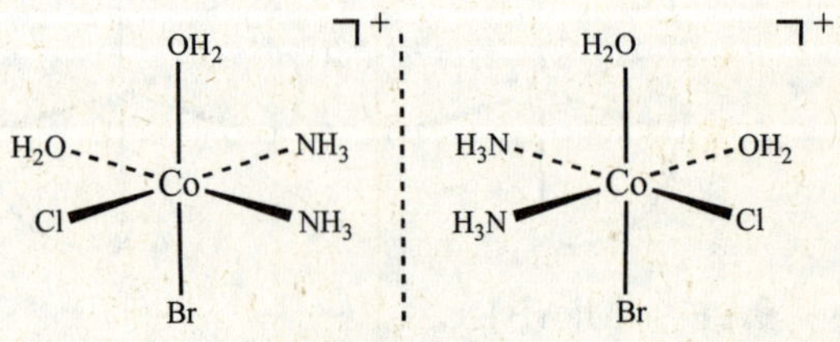 enantiomers

d. $Cr(H_2O)_3BrClI$

enantiomers

e. $[Pt(en)_2Cl_2]^{2+}$

cis enantiomers

trans

f. $[Cr(o\text{-phen})(NH_3)_2Cl_2]^+$

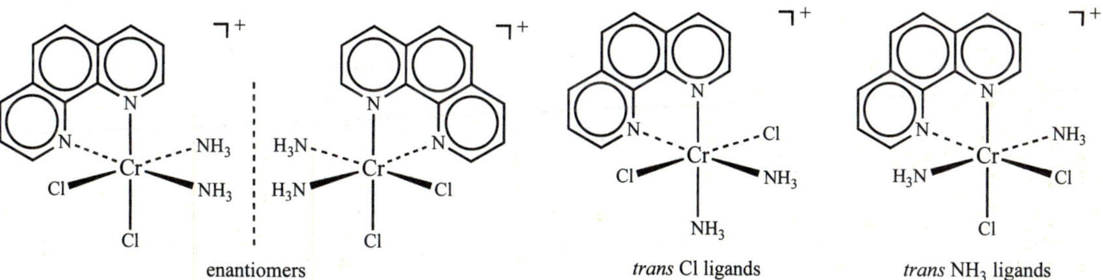

g. $[Pt(bipy)_2BrCl]^{2+}$

enantiomers

h. $Re(arphos)_2Br_2$ Abbreviating the bidentate ligands As⌒P:

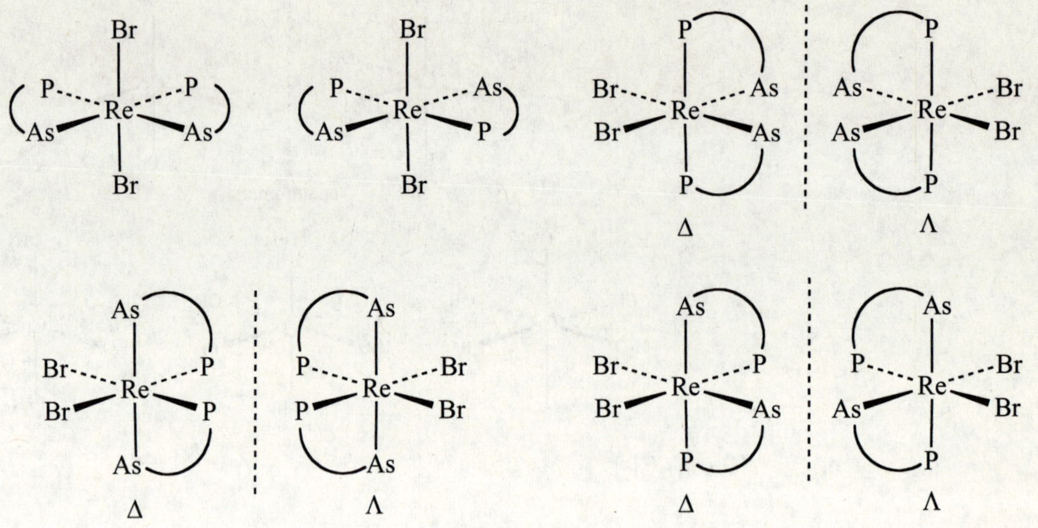

Δ Λ

Δ Λ Δ Λ

i. $Re(dien)Br_2Cl$

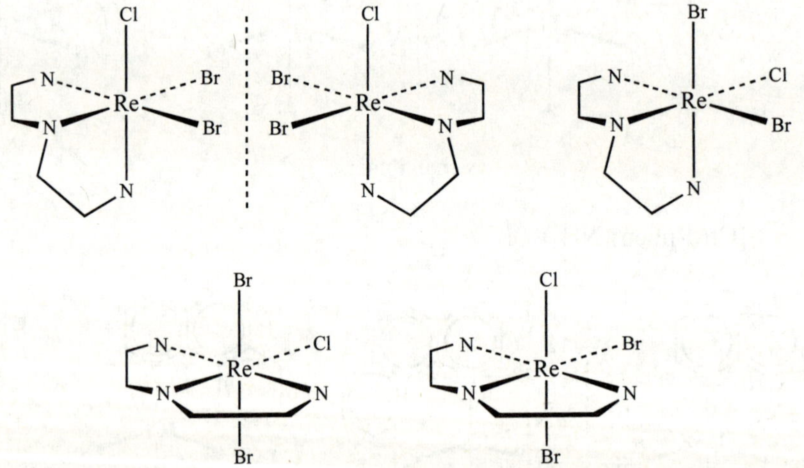

9.13 **a.** The "softer" phosphorus atom bonds preferentially to the soft metal Pd (see section 6.3.1).

b, c. Abbreviating the bidentate ligands N⌒P:

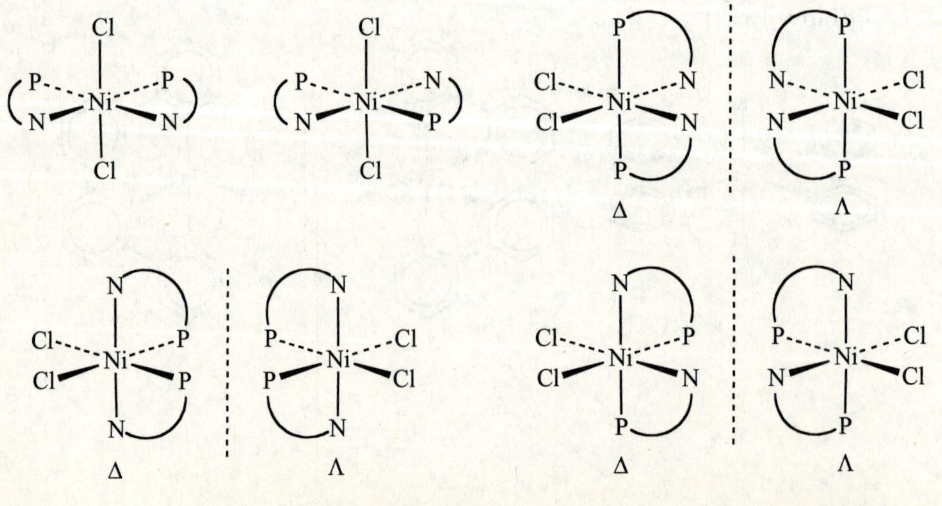

Δ Λ

Δ Λ Δ Λ

9.14 **a.** Abbreviating the bidentate ligands N͡P and O͡S:

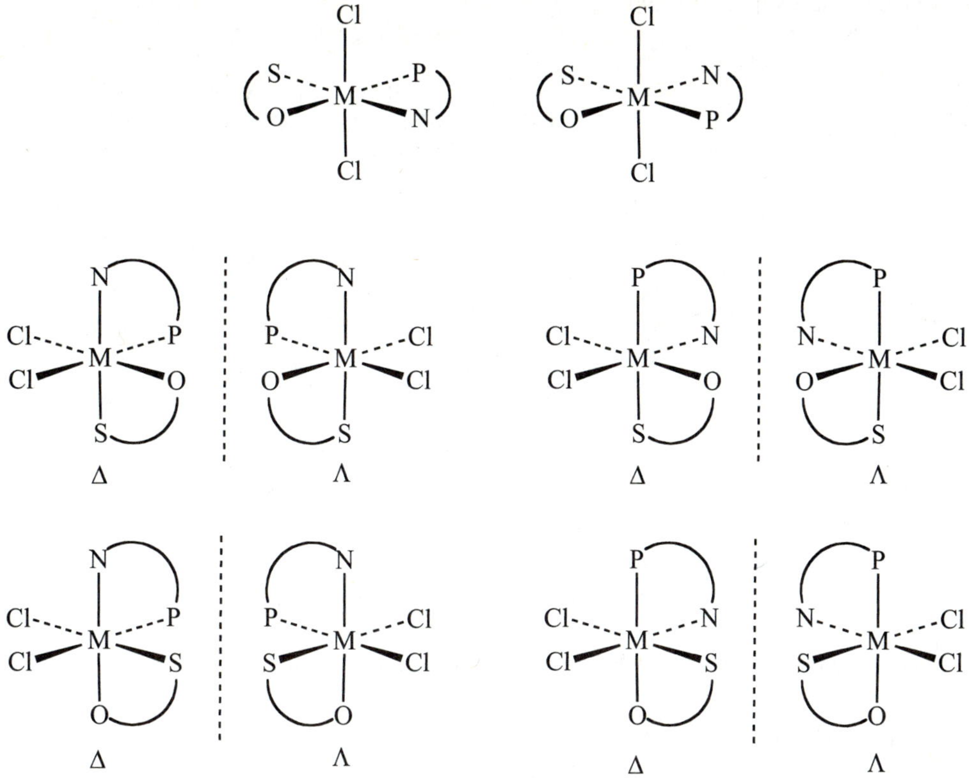

Δ Λ Δ Λ

Δ Λ Δ Λ

9.15 The single C–N stretching frequency indicates a *trans* structure for the cyanides (the symmetric stretch of the C–N bonds is not IR active), while the two C–O bands indicate a *cis* structure for the carbonyls (both the symmetric and antisymmetric C–O stretches are IR active). As a result, the bromo ligands are also *cis*.

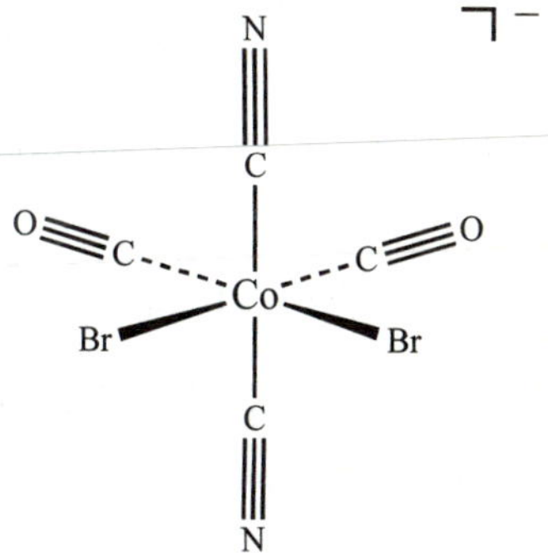

9.16 There are 18 isomers overall, twelve with the chelating ligand in a *mer* geometry and six with the chelating ligand in a *fac* geometry. All are enantiomers. They are all shown below, with dashed lines separating the enantiomers.

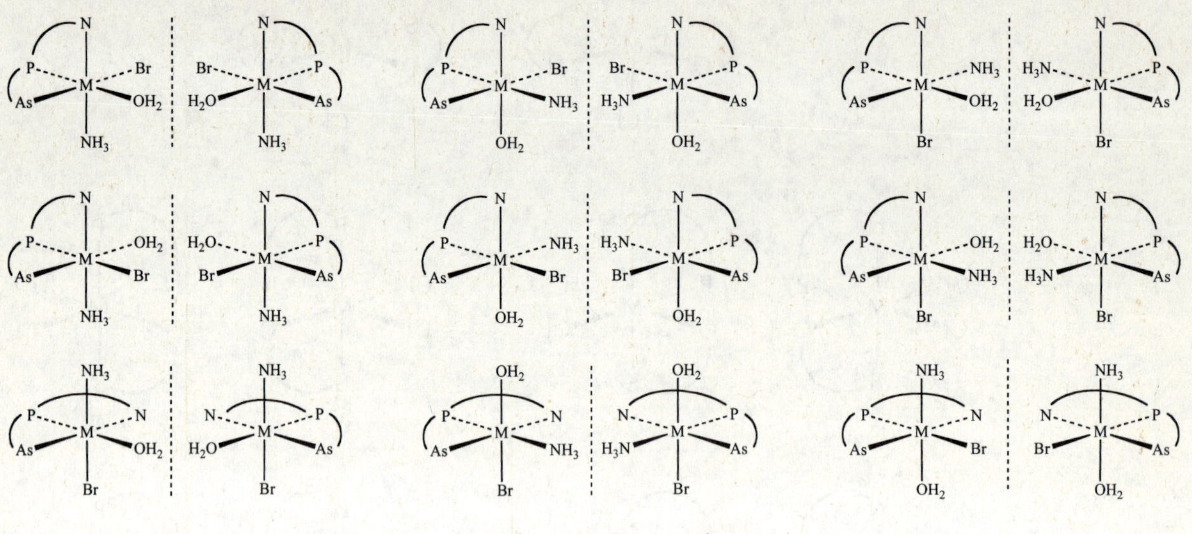

9.17 **a.** Δ **b.** Δ **c.** Λ **d.** Δ

9.18 All are chiral if the ring in **b** does not switch conformations.

9.19 18b δ 18c top ring: δ, bottom ring: λ

9.20 The ^{19}F doublet is from the two axial fluorines (split by the equatorial fluorine).

The ^{19}F triplet is from the equatorial fluorine (split by the two axial fluorines).

The two doubly bonded oxygens are equatorial, as expected from VSEPR considerations.

9.21 Examples include both cations and anions:

$[Cu(CN)_2]^-$, $[Cu_2(CN)_3]^-$, $[Cu_3(CN)_4]^-$, $[Cu_4(CN)_5]^-$, $[Cu_5(CN)_6]^-$

$[Cu_2(CN)]^+$, $[Cu_3(CN)_2]^+$, $[Cu_4(CN)_3]^+$, $[Cu_5(CN)_4]^+$, $[Cu_6(CN)_5]^+$

Based primarily on calculations (rather than experimental data), Dance et al. have proposed linear structures such as the following.:

$[Cu(CN)_2]^-$:	NC—Cu—CN
$[Cu_2(CN)_3]^-$:	NC—Cu—CN—Cu—CN
$[Cu_3(CN)_4]^-$:	NC—Cu—CN—Cu—CN—Cu—CN
$[Cu_2(CN)]^+$:	Cu—CN—Cu
$[Cu_3(CN)_2]^+$:	Cu—CN—Cu—CN—Cu
$[Cu_4(CN)_3]^+$:	Cu—CN—Cu—NC—Cu—CN—Cu

Where 2-coordinate copper appears in these ions, the geometry around the Cu is linear, as expected from VSEPR.

9.22 The bulky mesityl groups cause sufficient crowding that the phosphine ligands can show chirality (C_3 symmetry) and can be considered as similar to left handed (P_L) and right handed (P_R) propellers. If two P(mesityl)$_3$ phosphines are attached in a linear arrangement to a gold atom, three isomers are possible:

H_3C CH_3 H_3C mesityl

P_L—Au—P_L P_R—Au—P_R P_L—Au—P_R

(P_R—Au—P_L is equivalent to P_L—Au—P_R, as can best be seen by making models.) NMR data at low temperature support the presence of these isomers, which interconvert at higher temperatures.

9.23 The point group is D_{3h}. A representation Γ based on the nine 1s orbitals of the hydride ligands is:

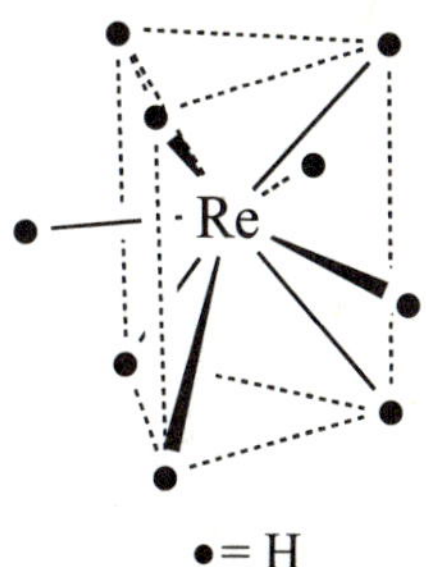

D_{3h}	E	$2C_3$	$3C_2$	σ_h	$2S_3$	$3\sigma_v$	
Γ	9	0	1	3	0	3	
A_1'	1	1	1	1	1	0	z^2
E'	2	−1	0	2	−1	0	$(x, y), (x^2-y^2, xy)$
A_2''	1	1	−1	−1	−1	1	z
E''	2	−1	0	−2	1	0	(xz, yz)

The representation Γ reduces to 2 A_1' + 2 E' + A_2'' + E''. Collectively these representations match all the functions for s (totally symmetric, matching A_1'), p, and d orbitals of Re, so all the s, p, and d orbitals of the metal have suitable symmetry for interaction. (The strength of these interactions will also depend on the match in energies between the rhenium orbitals and the 1s orbital of hydrogen.)

9.24

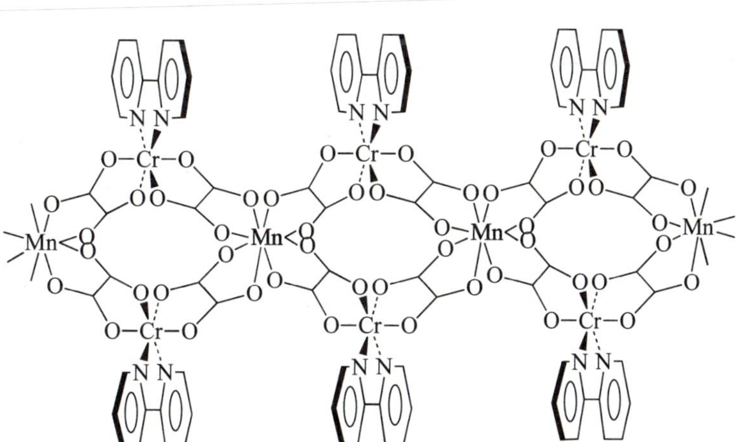

9.25

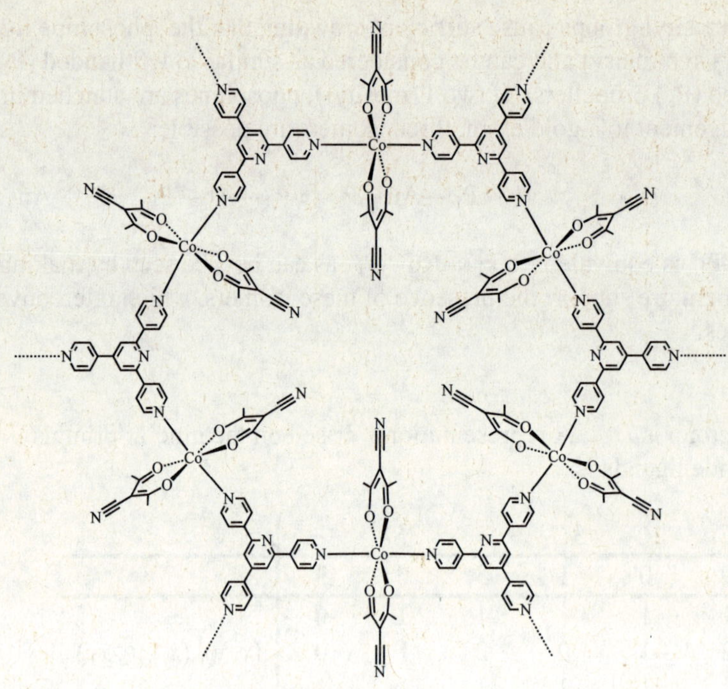

9.26 **a.** Cu(acacCN)$_2$: D_{2h} tpt: C_{2v}

b. C_6

CHAPTER 10: COORDINATION CHEMISTRY II: BONDING

10.1 **a.** Tetrahedral d^6, 4 unpaired electrons

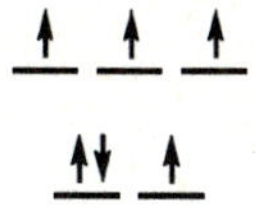

b. $[Co(H_2O)_6]^{2+}$, high spin octahedral d^7, 3 unpaired electrons

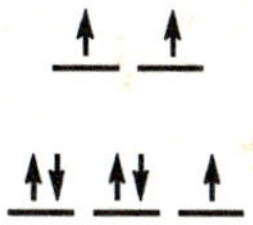

c. $[Cr(H_2O)_6]^{3+}$, octahedral d^3, 3 unpaired electrons

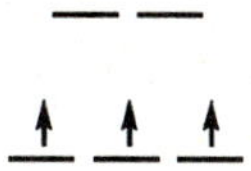

d. square planar d^7, 1 unpaired electron

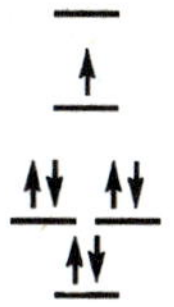

e. 5.1 BM $= \mu = \sqrt{n(n+2)}$; $n = 4.2 \approx 4$

10.2 **a.** $[\mathbf{M}(H_2O)_6]^{3+}$ with 1 unpaired electron. d^1: **M** = Ti

b. $[\mathbf{M}Br_4]^-$ with the maximum number of unpaired electrons (5).

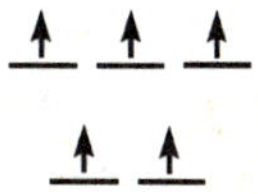

$\mathbf{M}^{3+}$ with 5 d electrons: **M** = Fe

c. Diamagnetic $[\mathbf{M}(CN)_6]^{3-}$. The strong field cyano ligand favors low spin. $\mathbf{M}^{3+}$ with 6 d electrons: **M** = Co

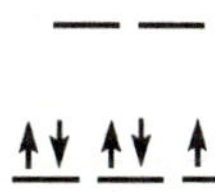

d. $[\mathbf{M}(H_2O)_6]^{2+}$ having LFSE $= -\frac{3}{5}\Delta_o$. Both high spin d^4 and d^9 have the correct Δ_o.

High spin d^4: **M** = Cr (0.6 Δ_o; – 0.4 Δ_o)

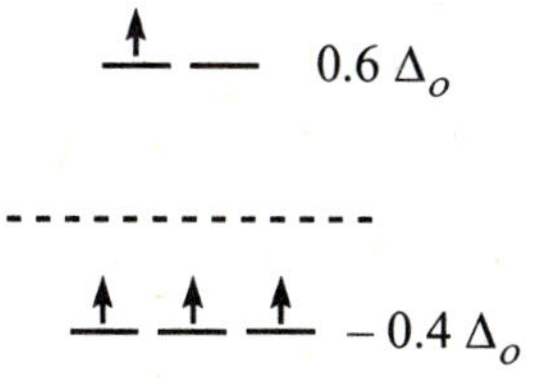

d^9: **M** = Cu (0.6 Δ_o; – 0.4 Δ_o)

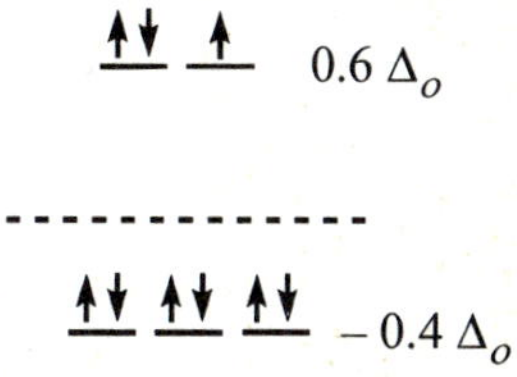

10.3 **a.** $K_3[\mathbf{M}(CN)_6]$ **M** is first row transition metal, 3 unpaired electrons.

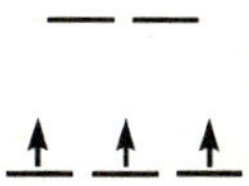

$\mathbf{M}^{3+}$ with 3 d electrons: **M** = Cr

b. $[\mathbf{M}(H_2O)_6]^{3+}$ **M** is second row transition metal, LFSE = $-2.4\ \Delta_o$. This can be achieved with a low spin d^6 configuration. $\mathbf{M}^{3+}$ with 6 d electrons: **M** = Rh

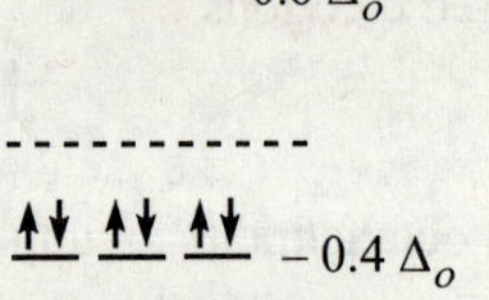

— — $0.6\ \Delta_o$

- - - - - - - - - - - - -

↑↓ ↑↓ ↑↓ $-0.4\ \Delta_o$

c. $[\mathbf{M}Cl_4]^-$ **M** is first row transition metal, 5 unpaired electrons. $\mathbf{M}^{3+}$ with 5 d electrons: Fe

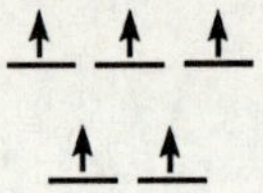

↑ ↑ ↑

↑ ↑

d. $\mathbf{M}Cl_2(NH_3)_2$ **M** is third row d^8 transition metal; two **M**–Cl stretching bands in IR. Second and third row d^8 complexes are often square planar. The presence of two **M**–Cl stretching bands implies *cis* geometry of the chloro ligands.

$\mathbf{M}^{2+}$ with 8 d electrons: **M** = Pt

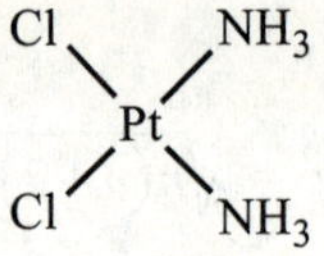

Cl, NH_3 / Pt / Cl, NH_3

10.4 Simple σ angular overlap calculations for d^8 and d^9 ions show no energy difference between D_{4h} and O_h (Both d^8 geometries have energies of $-3\ e_\sigma$; both d^9 geometries have energies of $-6\ e_\sigma$). In general, stability constants decrease as more ligands are added, so the sequence for nickel is the common one. The huge drop in stability constant between the second and third ethylenediamine on Cu^{2+} is a result of the d^9 Jahn-Teller effect. The first two en ligands add in a square-planar geometry, with water molecules in the axial positions, and the difference between the ligands allows for the Jahn-Teller distortion. Adding a third en ligand requires a shift in geometry and pushes the bond distances closer together. This is counter to the Jahn-Teller distortion, and as a result the third addition is much less favorable than the first two.

10.5 $[\mathbf{M}(H_2O)_6]^{2+}$ **M** is first row transition metal, μ = 3.9 BM. The magnetic moment implies 3 unpaired electrons, which would give rise to $\mu = \sqrt{3(3+5)}$ = 3.9 BM. There are two possibilities, d^3 and d^7:

$\mathbf{M}^{2+}$, d^3 : **M** = V

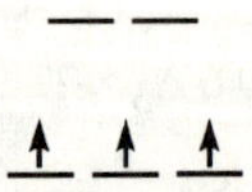

— —

↑ ↑ ↑

$\mathbf{M}^{2+}$, d^7: **M** = Co

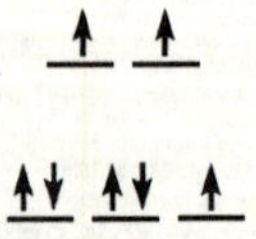

↑ ↑

↑↓ ↑↓ ↑

10.6 **a.** $[Cr(H_2O)_6]^{2+}$ $n = 4$
$\mu = \sqrt{4(6)} = 4.9\ \mu_B$

b. $[Cr(CN)_6]^{4-}$ $n = 2$
$\mu = \sqrt{2(4)} = 2.8\ \mu_B$

c. $[FeCl_4]^-$ $n = 5$
$\mu = \sqrt{5(7)} = 5.9\ \mu_B$

d. $[Fe(CN)_6]^{3-}$ $n = 1$
$\mu = \sqrt{1(3)} = 1.7\ \mu_B$

e. $[Ni(H_2O)_6]^{2+}$ $n = 2$
$\mu = \sqrt{2(4)} = 2.8\ \mu_B$

f. $[Cu(en)_2(H_2O)_2]^{2+}$ $n = 1$
$\mu = \sqrt{1(3)} = 1.7\ \mu_B$

10.7 $Fe(H_2O)_4(CN)_2$ is really $[Fe(H_2O)_6]_2[Fe(CN)_6]$, all containing Fe(II). $[Fe(H_2O)_6]^{2+}$ is high spin d^6, with $\mu = 4.9\ \mu_B$; $[Fe(CN)_6]^{4-}$ is low spin d^6, with $\mu = 0\ \mu_B$. The average value is then $2 \times 4.9/3 = 3.3\ \mu_B$.

2.67 unpaired electrons gives $\mu = \sqrt{2.67 \times 4.67} = 3.53\ \mu_B$.

10.8 Co(II) is d^7. In tetrahedral complexes, it is high spin and has 3 unpaired electrons; in octahedral complexes, it is typically high spin and also has 3 unpaired electrons; in square planar complexes, it has 1 unpaired electron. The magnetic moments can be calculated as $\mu = \sqrt{n(n+2)} = 3.9$, 3.9, and 1.7 μ_B, respectively.

10.9 For the red compounds (Me and Et at high temperatures, Pr, pip, and pyr at all temperatures), the larger magnetic moment indicates approximately 5 unpaired electrons, appropriate for high-spin Fe(III) species. At low temperatures for the Me and Et compounds, the magnetic moment indicates 3 to 4 unpaired electrons, apparently an average value indicating an equilibrium mixture of high and low spin species. The low spin octahedral complexes would have one unpaired electron. Increasing the size of the R groups changes the structure enough that it is locked into high-spin species at all temperatures.

10.10 Both $[M(H_2O)_6]^{2+}$ and $[M(NH_3)_6]^{2+}$ should show the double-humped curve of Figure 10.13, with larger values for the NH_3 compounds. Therefore, the difference between these curves still shows some of the same shape.

10.11 These can be verified using the approach introduced in Chapters 4 and 5, in which a character of +1 is counted for each operation that leaves a vector unchanged, a character of –1 for each operation that leaves a vector in its position but with the direction reversed, and a character of 0 for other situations.

10.12 The e_σ column gives d orbital energies for complexes involving σ donor ligands only; the Total column gives energies for complexes of ligands that act as both σ donors and π acceptors:

a. ML_2, using positions 1 and 6:

	e_σ	e_π	Total
z^2	2	0	$2e_\sigma$
x^2-y^2	0	0	0
xy	0	0	0
xz	0	–2	$-2\ e_\pi$
yz	0	–2	$-2\ e_\pi$

b. ML_3, using positions 2, 11, 12:

	e_σ	e_π	Total
z^2	0.75	0	$0.75e_\sigma$
x^2-y^2	1.125	–1.5	$1.125\ e_\sigma - 1.5\ e_\pi$
xy	1.125	1.5	$1.125\ e_\sigma - 1.5\ e_\pi$
xz	0	–1.5	$-1.5\ e_\pi$
yz	0	–1.5	$-1.5\ e_\pi$

c. ML_5, C_{4v}, using positions 1, 2, 3, 4, 5:

	e_σ	e_π	Total
z^2	2	0	$2e_\sigma$
$x^2–y^2$	3	0	$3e_\sigma$
xy	0	–4	$–4e_\pi$
xz	0	–3	$–3e_\pi$
yz	0	–3	$–3e_\pi$

d. ML_5, D_{3h}, using positions 1, 2, 6, 11, 12:

	e_σ	e_π	Total
z^2	2.75	0	$2.75e_\sigma$
$x^2–y^2$	1.125	–1.5	$1.125\ e_\sigma – 1.5\ e_\pi$
xy	1.125	–1.5	$1.125\ e_\sigma – 1.5\ e_\pi$
xz	0	–3.5	$–3.5\ e_\pi$
yz	0	–3.5	$–3.5\ e_\pi$

e. ML_8, cube, positions 7, 8, 9, 10, doubled for the other four corners:

	e_σ	e_π	Total
z^2	0	–5.33	$–5.33\ e_\pi$
$x^2–y^2$	0	–5.33	$–5.33\ e_\pi$
xy	2.67	–1.78	$2.67\ e_\sigma – 1.78\ e_\pi$
xz	2.67	–1.78	$2.67\ e_\sigma – 1.78\ e_\pi$
yz	2.67	–1.78	$2.67\ e_\sigma – 1.78\ e_\pi$

10.13

Metal d orbitals, NH_3 influence

	e_σ	e_π	Total
z^2	1	0	$1e_\sigma$
$x^2–y^2$	3	0	$3e_\sigma$
xy	0	0	0
xz	0	0	0
yz	0	0	0

Metal d orbitals, $Cl^–$ influence

	e_σ	e_π	Total
z^2	2	0	$2e_\sigma$
$x^2–y^2$	0	0	0
xy	0	0	0
xz	0	2	$2\ e_\pi$
yz	0	2	$2\ e_\pi$

Ligand NH_3:

	e_σ	Total
1	0	0
2	–1	$–1e_\sigma$
3	–1	$–1e_\sigma$
4	–1	$–1e_\sigma$
5	–1	$–1e_\sigma$
6	0	0

Ligand $Cl^–$:

	e_σ	e_π	Total
1	–1	–2	$–1e_\sigma – 2\ e_\pi$
2	0	0	0
3	0	0	0
4	0	0	0
5	0	0	0
6	–1	–2	$–1e_\sigma – 2\ e_\pi$

Overall energy = $–8\ e_\sigma(NH_3) – 4\ e_\sigma(Cl) – 8\ e_\pi(Cl) + 4\ e_\pi(Cl) = –8\ e_\sigma(NH_3) – 4\ e_\sigma(Cl) – 4\ e_\pi(Cl)$
The metal electrons are still unpaired, one in the d_{xy} orbital and on each in the d_{xz} and d_{yz}, raised by π interaction with $Cl^–$. Four of the ligand orbitals are lowered by $e_\sigma(NH_3)$ and two are lowered by $e_\sigma(Cl) + 2\ e_\pi(Cl)$; each contains a pair of ligand electrons. Since the magnitudes of the changes in energy are not known, the order of the ligand orbital energies is uncertain.

10.14 **a.** The σ bonding energies are given in problem 10.12d.

b. For an axial L′, the π bonding would affect only the d_{xz} and d_{yz} orbitals, with energies stabilized by $1e_\pi$ for each. In the equatorial position 2, L′ lowers the energy of the *xy* and *yz* orbitals by $1e_\pi$ for each. In positions 11 or 12, the $d_{x^2-y^2}$ and d_{yz} orbitals are lowered by 0.75 e_π, and d_{xy} and d_{xz} are lowered by $0.25e_\pi$. The average effect is that d_{z^2} is unaffected, and the other four orbitals are stabilized by 0.50 e_π.

c. Stability depends on the number of *d* electrons; 1 through 7 electrons result in a stabilization of the axial isomer, 8 through 10 electrons result in no difference between the isomers.

10.15 The angular overlap differences (calculated in problem 10.12, c and d) between square pyramidal and trigonal bipyramidal structures are not large, but generally favor square pyramidal structures. Other factors (such as VSEPR) also contribute.

Sigma Interactions Only	High spin		Low spin	
Number of *d* electrons	Square pyramidal	Trigonal bipyramidal	Square pyramidal	Trigonal bipyramidal
1	0	0	0	0
2	0	0	0	0
3	0	1.125 e_σ	0	0
4	2 e_σ	2.25 e_σ	0	0
5	5 e_σ	5 e_σ	0	1.125 e_σ
6	5 e_σ	5 e_σ	0	2.25 e_σ
7	5 e_σ	5 e_σ	2 e_σ	3.375 e_σ
8	5 e_σ	6.125 e_σ	5 e_σ	4.50 e_σ
9	7 e_σ	7.25 e_σ	7 e_σ	7.25 e_σ
10	10 e_σ	10 e_σ	10 e_σ	10 e_σ

Sigma and Pi Interactions	High Spin		Low Spin	
Number of *d* electrons	Square pyramidal	Trigonal bipyramidal	Square pyramidal	Trigonal bypyramidal
1	$-4\ e_\pi$	$-4\ e_\pi$	$-4\ e_\pi$	$-3.5\ e_\pi$
2	$-7\ e_\pi$	$-7\ e_\pi$	$-8\ e_\pi$	$-7\ e_\pi$
3	$-10\ e_\pi$	$1.125\ e_\sigma - 8.5\ e_\pi$	$-11\ e_\pi$	$-10.5\ e_\pi$
4	$2\ e_\sigma - 10\ e_\pi$	$2.25\ e_\sigma - 10\ e_\pi$	$-14\ e_\pi$	$-14\ e_\pi$
5	$5\ e_\sigma - 10\ e_\pi$	$5\ e_\sigma - 13.5\ e_\pi$	$-17\ e_\pi$	$1.125\ e_\sigma - 15.5\ e_\pi$
6	$5\ e_\sigma - 14\ e_\pi$	$5\ e_\sigma - 17e_\pi$	$-20\ e_\pi$	$2.25\ e_\sigma - 17\ e_\pi$
7	$5\ e_\sigma - 17\ e_\pi$	$5\ e_\sigma - 17\ e_\pi$	$2\ e_\sigma - 20\ e_\pi$	$3.375\ e_\sigma - 18.5\ e_\pi$
8	$5\ e_\sigma - 20\ e_\pi$	$6.125\ e_\sigma - 18.5\ e_\pi$	$5\ e_\sigma - 20\ e_\pi$	$4.50\ e_\sigma - 20\ e_\pi$
9	$7\ e_\sigma - 20\ e_\pi$	$7.25\ e_\sigma - 20\ e_\pi$	$7\ e_\sigma - 20\ e_\pi$	$7.25\ e_\sigma - 20\ e_\pi$
10	$10\ e_\sigma - 20\ e_\pi$	$10\ e_\sigma - 20\ e_\pi$	$10\ e_\sigma - 20\ e_\pi$	$10\ e_\sigma - 20\ e_\pi$

10.16 **a.** Seesaw, using positions 1, 6, 11, and 12 (σ donor only)

Position	z^2	x^2-y^2	*xy*	*xz*	*yz*
1	1	0	0	0	0
6	1	0	0	0	0
11	1/4	3/16	9/16	0	0
12	1/4	3/16	9/16	0	0
Total	2.5	0.375	1.125	0	0

Trigonal pyramidal, using positions 1, 2, 11, and 12 (σ donor only)

Position	z^2	$x^2 - y^2$	xy	xz	yz
1	1	0	0	0	0
2	1/4	3/4	0	0	0
11	1/4	3/16	9/16	0	0
12	1/4	3/16	9/16	0	0
Total	1.75	1.125	1.125	0	0

b.

Number of d Electrons	Energies of d Configurations (in units of e_σ) Seesaw Low Spin	Seesaw High Spin	Trigonal Pyramidal Low Spin	Trigonal Pyramidal High Spin
1	0	0	0	0
2	0	0	0	0
3	0	0.375	0	1.125
4	0	1.5	0	2.25
5	0.375	4	1.125	4
6	0.750	4	2.25	4
7	1.875	4	3.375	4
8	3	4.375	4.5	5.125
9	5.5	5.5	6.25	6.25
10	8	8	8	8

c. The seesaw geometry is favored for d^5- d^9 low spin and d^3, d^4, d^8, and d^9 high spin configurations. For other configurations, both geometries have identical energies according to this approach. (The values shown are for the d orbitals only; each interaction also stabilizes a ligand orbital by e_σ.)

10.17 a. The new positions (13 and 14 below) are opposite 11 and 12, and have the same values in the table. z^2 is affected most strongly, since two ligands are along the z axis; $x^2 - y^2$ and xy are also strongly influenced, since they are in the plane of the six ligands. xz and yz are not changed, since they miss the ligands in all directions.

Position	z^2	$x^2 - y^2$	xy	xz	yz
1	1	0	0	0	0
2	1/4	3/4	0	0	0
6	1	0	0	0	0
11	1/4	3/16	9/16	0	0
12	1/4	3/16	9/16	0	0
4	1/4	3/4	0	0	0
13	1/4	3/16	9/16	0	0
14	1/4	3/16	9/16	0	0
Total	3.5	2.25	2.25	0	0

b. From the table below, we can see that $z^2 = A_{1g}$, $(x^2 - y^2, xy) = E_{2g}$, and $(xz, yz) = E_{1g}$

LFSE Δ_o LFSE Difference

D_{6h}	E	$2C_6$	$2C_3$	C_2	$3C_2'$	$3C_2''$	i	$2S_3$	$2S_6$	σ_h	$3\sigma_d$	$3\sigma_v$		
Γ	8	2	2	2	2	0	0	0	0	6	2	4		
A_{1g}	1	1	1	1	1	1	1	1	1	1	1	1		z^2
A_{2g}	1	1	1	1	−1	−1	1	1	1	1	−1	−1		
B_{1g}	1	−1	1	−1	1	−1	1	−1	1	−1	1	−1		
B_{2g}	1	−1	1	−1	−1	1	1	−1	1	−1	−1	1		
E_{1g}	2	1	−1	−2	0	0	2	1	−1	−2	0	0		(xz, yz)
E_{2g}	2	−1	−1	2	0	0	2	−1	−1	2	0	0		(x^2-y^2, xy)
A_{1u}	1	1	1	1	1	1	−1	−1	−1	−1	−1	−1		
A_{2u}	1	1	1	1	−1	−1	−1	−1	−1	−1	1	1	z	
B_{1u}	1	−1	1	−1	1	−1	−1	1	−1	1	−1	1		
B_{2u}	1	−1	1	−1	−1	1	−1	1	−1	1	1	−1		
E_{1u}	2	1	−1	−2	0	0	−2	−1	1	2	0	0	(x, y)	
E_{2u}	2	−1	−1	2	0	0	−2	1	1	−2	0	0		

$\Gamma = 2A_{1g} + E_{2g} + A_{2u} + B_{1u} + E_{1u}$

$s, z^2 \quad (x^2-y^2, xy) \quad z \quad (x, y)$

c.

Position	z^2	x^2-y^2	xy	xz	yz
1	0	0	0	1	1
2	0	0	1	1	0
6	0	0	0	1	1
11	0	3/4	1/4	1/4	3/4
12	0	3/4	1/4	1/4	3/4
4	0	0	1	1	0
13	0	3/4	1/4	1/4	3/4
14	0	3/4	1/4	1/4	3/4
Total	0	$3\,e_\pi$	$3\,e_\pi$	$5\,e_\pi$	$5\,e_\pi$
Overall	$3.5\,e_\sigma$	$2.25\,e_\sigma + 3\,e_\pi$	$2.25\,e_\sigma + 3\,e_\pi$	$5\,e_\pi$	$5\,e_\pi$

Because the Jahn-Teller effect is larger when the lowest level is unevenly populated, d^1 and d^3 would have the largest effect. The location of $d_{x^2-y^2}$ and d_{xy} relative to the uncoordinated metal ion is uncertain; it depends on the magnitudes of e_σ and e_π.

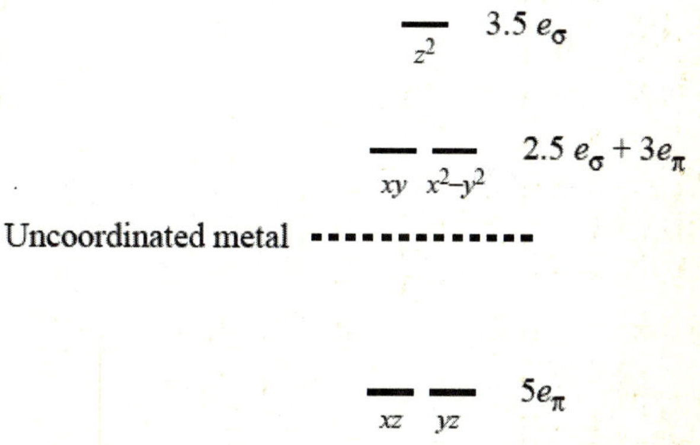

z^2 — $3.5\,e_\sigma$

xy, x^2-y^2 — $2.5\,e_\sigma + 3e_\pi$

Uncoordinated metal

xz, yz — $5e_\pi$

10.18

$[Co(H_2O)_6]^{3+}$	d^6 (ls)	$-2.4\Delta_o$	18,000	–43,200	35,440
$[Co(H_2O)_6]^{2+}$	d^7 (hs)	$-0.8\Delta_o$	9,700	–7,760	
$[Co(NH_3)_6]^{3+}$	d^6 (ls)	$-2.4\Delta_o$	24,000	–57,600	49,440
$[Co(NH_3)_6]^{2+}$	d^7 (hs)	$-0.8\Delta_o$	10,200	–8,160	

The ammine Co(III) complex is considerably more stable and is less easily reduced, with the difference primarily in the 3+ species. In addition, the metal in $[Co(NH_3)_6]^{3+}$ is surrounded by ammonia molecules, which are more difficult to oxidize than water. This makes transfer of electrons through the ligand more difficult for the ammine complexes.

10.19 Cl^- has the lowest Δ_o value and fairly good π donor properties that reduce Δ_o. F^- is next, with less π donor ability. Water has very small π donor ability (only one lone pair not involved in σ bonding), and ammonia and en have neither π donor nor acceptor ability (no lone pairs, antibonding orbitals with the wrong shapes and energies for bonding). CN^- has good π acceptor properties, making Δ_o largest for this ligand.

10.20 Ammonia is a stronger field ligand than water. It is a stronger Lewis base (σ donor) than water. In addition, water has a lone pair that can act as a π donor. These factors are unfavorable enough for water that the less electronegative nitrogen on ammonia is a better donor atom. In the halide ions, all have the same valence electronic structure, so the electronegativity is the determining factor in ligand field strength. Fluoride is also a stronger base than the other halide ions.

10.21 **a.** Compression moves d_{z^2} up in energy and lowers the ligand energies of positions 1 and 6.

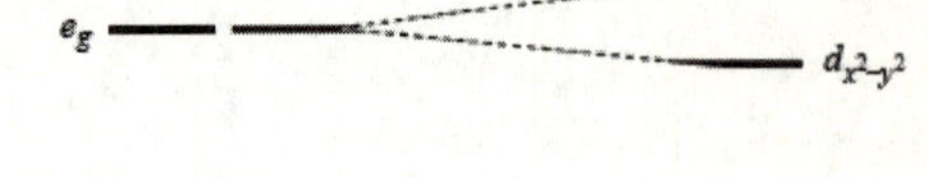

b. Stretching reverses the changes. In the limit of a square planar structure, d_{z^2} is affected only through interactions with the ring in the *xy* plane.

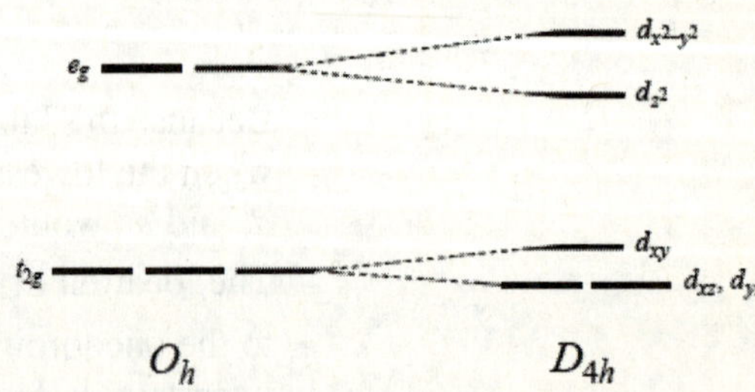

10.22 Cr^{3+} has three singly occupied t_{2g} orbitals and two empty e_g orbitals. As a result, there is no Jahn-Teller distortion in its complexes. Mn^{3+} has one electron in the e_g orbitals, so there is Jahn-Teller distortion in its complexes.

10.23 a.

	n	μ	Term	LFSE (in Δ_o)
$[Co(CO)_4]^-$	0	0	^1S	0
$[Cr(CN)_6]^{4-}$	2	2.8	^3H	–1.6
$[Fe(H_2O)_6]^{3+}$	5	5.9	^6S	0
$[Co(NO_2)_6]^{4-}$	3	3.9	^4F	–0.8
$[Co(NH_3)_6]^{3+}$	0	0	^1I	–2.4
$[MnO_4]^-$	0	0	^1S	0
$[Cu(H_2O)_6]^{2+}$	1	1.7	^2D	–0.6

b. The two tetrahedral ions ($[Co(CO)_4]^-$ and $[MnO_4]^-$) have zero LFSE (10 or 0 *d* electrons, respectively) and can have π bonding, with CO as an acceptor and O^{2-} as donor. With the exception of $[Fe(H_2O)_6]^{3+}$, the others have LFSE values that favor octrahedral structures. $[Fe(H_2O)_6]^{3+}$ has LFSE = 0 for either octrahedral or tetrahedral shapes, but water is a slight π donor and Fe(III) is only a moderately good π acceptor. As a result, electrostatics favors six ligands.

c. For Co^{2+}: High-spin octahedral d^7 has LFSE = $-0.8\Delta_o$
Tetrahedral d^7 has LFSE = $-1.2\Delta_t = -0.53\Delta_o$

For Ni^{2+}: High-spin octahedral d^8 has LFSE = $-1.2\Delta_o$
Tetrahedral d^8 has LFSE = $-0.6\Delta_t = -0.27\Delta_o$

Co(II) (d^7) has only $-0.27\Delta_o$ favoring the octahedral shape, while Ni(II) (d^8) has $-0.93\Delta_o$. Therefore, Co(II) compounds are more likely to be tetrahedral than are Ni(II) compounds.

10.24 a.

	O_h LFSE	T_d LFSE	$O_h - T_d$
d^1	$-0.4\Delta_o$	$-0.6\Delta_t = -0.27\Delta_o$	$-0.13\Delta_o$
d^2	$-0.8\Delta_o$	$-1.2\Delta_t = -0.53\Delta_o$	$-0.27\Delta_o$
d^3	$-1.2\Delta_o$	$-0.8\Delta_t = -0.36\Delta_o$	$-0.84\Delta_o$
d^4	$-0.6\Delta_o$	$-0.4\Delta_t = -0.18\Delta_o$	$-0.42\Delta_o$
d^5	$0\Delta_o$	$0\Delta_t = 0\Delta_o$	$0\Delta_o$
d^6	$-0.4\Delta_o$	$-0.6\Delta_t = -0.27\Delta_o$	$-0.13\Delta_o$
d^7	$-0.8\Delta_o$	$-1.2\Delta_t = -0.53\Delta_o$	$-0.27\Delta_o$
d^8	$-1.2\Delta_o$	$-0.8\Delta_t = -0.36\Delta_o$	$-0.84\Delta_o$
d^9	$-0.6\Delta_o$	$-0.4\Delta_t = -0.18\Delta_o$	$-0.42\Delta_o$
d^{10}	$0\Delta_o$	$0\Delta_t = 0\Delta_o$	$0\Delta_o$

The differences in *d* orbital energies always favor octahedral complexes, with the order of the absolute values $d^5 < d^6 < d^7 < d^9 < d^8$ and $d^1 < d^2 < d^4 < d^3$. For example, this order predicts the relative stability of tetrahedral complexes as Mn > Fe > Co > Cu > Ni. Fe is out of order, Co and Ni are in the correct order.

b. Angular overlap calculations result in $-e_\sigma$ for each ligand electron, or $-12\ e_\sigma$ for an octahedral complex and $-8\ e_\sigma$ for a tetrahedral complex. The metal electrons have the energies shown in the following table.

	O_h	T_d	$O_h - T_d$
d^1	$-12e_\sigma$	$-8e_\sigma$	$-4e_\sigma$
d^2	$-12e_\sigma$	$-8e_\sigma$	$-4e_\sigma$
d^3	$-12e_\sigma$	$-6.67e_\sigma$	$-5.33e_\sigma$
d^4	$-9e_\sigma$	$-5.33e_\sigma$	$-3.67e_\sigma$
d^5	$-6e_\sigma$	$-4e_\sigma$	$-2e_\sigma$
d^6	$-6e_\sigma$	$-4e_\sigma$	$-2e_\sigma$
d^7	$-6e_\sigma$	$-4e_\sigma$	$-2e_\sigma$
d^8	$-6e_\sigma$	$-2.67e_\sigma$	$-3.33e_\sigma$
d^9	$-3e_\sigma$	$-1.33e_\sigma$	$-1.67e_\sigma$
d^{10}	$0e_\sigma$	$0e_\sigma$	$0e_\sigma$

Again, all cases favor octahedral geometry, but here the relative order for tetrahedral geometry is $d^9 > d^6 = d^7 > d^8$. Nickel and cobalt are in the correct order, but iron is equal to cobalt rather than less likely to have tetrahedral geometry.

10.25 The energy levels of the square planar *d* orbitals are shown in Figure 10.15; those of the octahedral orbitals are shown in Figure 10.6. The seventh, eighth, and ninth *d* electrons in an octahedral complex go into the highest orbitals, raising the total energy of the complex. In these cases, the energy of the square planar complex may be more favorable even though the total ligand electron energy is less than in octahedral complexes because there are only four ligands rather than six.

10.26 a. Square pyramidal complexes have C_{4v} symmetry.

C_{4v}	E	$2C_4$	C_2	$2\sigma_v$	$2\sigma_d$	
Γ	5	1	1	3	1	
A_1	1	1	1	1	1	z, z^2
B_1	1	-1	1	1	-1	x^2-y^2
E	2	0	-2	0	0	$(x, y), (xz, yz)$

The z^2 and x^2–y^2 orbitals are the major *d* orbitals used in the bonding, so the *d* orbital energies are as shown at right. The z^2 orbital is less involved because it is directed primarily toward one ligand, and the x^2–y^2 orbital is directed toward four ligands.

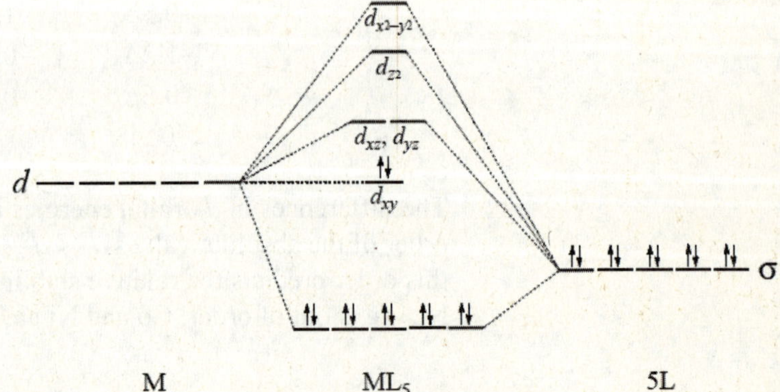

b. Pentagonal bipyramidal complexes have D_{5h} symmetry.

D_{5h}	E	$2C_5$	$2C_5^2$	$5C_2$	σ_h	$2S_5$	$2S_5^3$	$5\sigma_v$	
Γ	7	2	2	1	5	0	0	3	
A_1'	1	1	1	1	1	1	1	1	z^2
E_1'	2	2 cos 72°	2 cos 144°	0	2	2 cos 144°	2 cos 72°	0	(x, y)
E_2'	2	2 cos 144°	2 cos 72°	0	2	2 cos 72°	2 cos 144°	0	$(x^2 - y^2, xy)$
A_2''	1	1	1	–1	–1	–1	–1	1	z

The representation Γ reduces to $2\,A_1' + E_1' + E_2' + A_2''$. The d orbitals are in three sets, including a nonbonding pair (d_{xz} and d_{yz}) in addition to the A_1' and degenerate E_2'.

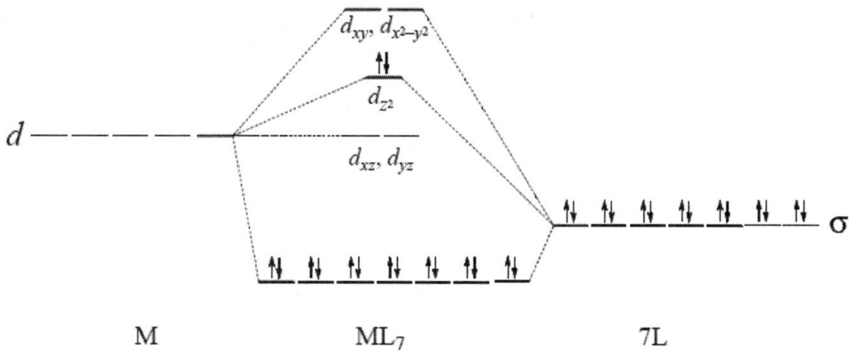

10.27 a. The MO diagram is similar to that for CO, shown in Figure 5.14, with the atomic orbitals of fluorine significantly lower than the matching orbitals of nitrogen. The NF molecule has two more electrons than CO; these occupy π^* orbitals and have parallel spins.

b. Because its highest occupied sigma orbital (labeled $3a_1$ for CO) is significantly concentrated on the nitrogen, NF should act as a strong σ donor. In addition, because the π^* orbitals are also strongly concentrated on nitrogen, NF should act as a π acceptor at nitrogen. Because the π^* orbitals are singly occupied, NF is likely to be a weaker π acceptor than CO, whose π^* orbitals are empty. Overall, NF is likely to be relatively high in the spectrochemical series, but not as high as CO.

10.28 a. When these compounds are oxidized, in the positively charged products there is less π acceptance by the CO ligands than in the neutral compounds. Because CO is acting less as a π acceptor in the products, the products have less donation into their π^* orbitals, which are antibonding with respect to the C–O bonds. This means that the oxidation products have stronger and shorter C–O bonds than the reactants. In the reference cited, the C–O bonds are calculated to be shortened by 0.014 to 0.018 Å in the PH_3 complex and by 0.015 to 0.020 Å in the NH_3 complex.

b. In the phosphine compound, oxidation reduces the π acceptance by the phosphine (as it reduces the π acceptance by the carbonyls), and there is weaker Cr–P bonding. Therefore, the Cr–P bond is longer in the product (by 0.094 Å).

In the ammine compound, the NH_3 is not a π acceptor; it is a σ donor only. Because oxidation increases the positive charge on the metal, there is stronger attraction between the metal and the sigma-donating NH_3, shortening the Cr–N bond (by 0.050 Å).

10.29 **a.** Group orbitals:

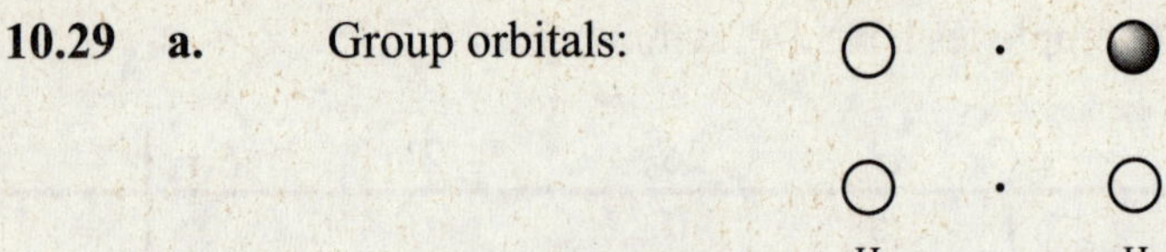

b. Group orbital–central atom interactions:

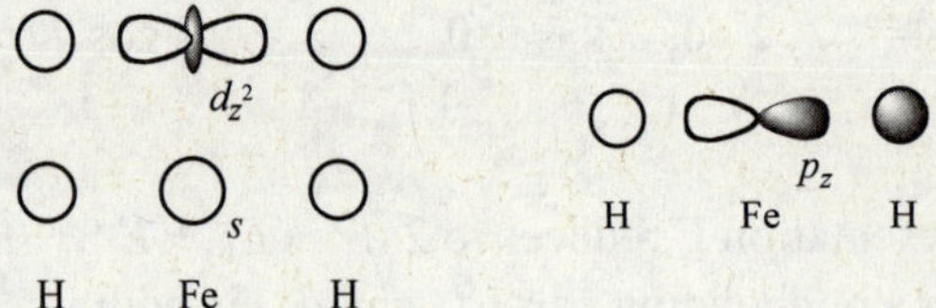

c. The hydrogen orbitals (potential energy = –13.6 eV) are likely to interact more strongly with iron orbitals that have a better energy match, the 3*d* (–11.7 eV) and 4*s* (–7.9 eV). The 3*p* orbitals of iron are likely to have very low energy (ca. –30 eV) and not to interact significantly with the hydrogen orbitals. (Potential energies for transition metal orbitals can be found in J. B. Mann, T. L. Meek, E. T. Knight, J. F. Capitani, and L. C. Allen, *J. Am. Chem. Soc.*, **2000**, *122*, 5132.)

10.30 **a.** D_{4h}

b. **1.** Because a deuterium atom has only its 1*s* electron to participate in bonding, there are only three platinum group orbitals that can potentially interact with deuterium:

Group Orbitals: Group Orbital–D Interactions

d_{z^2}

p_z

s

Pt D Pt Pt D Pt

2. In each case, sigma interactions could occur, as shown above.

3. The strongest interactions are most likely between the d_{z^2} orbitals of Pt and the 1s orbital of D. Lobes of the $5d_{z^2}$ orbitals point toward the D, and their energy (–10.4 eV) is a good match for the valence orbital potential energy of hydrogen (–13.6 eV). Both bonding and antibonding molecular orbitals would be formed:

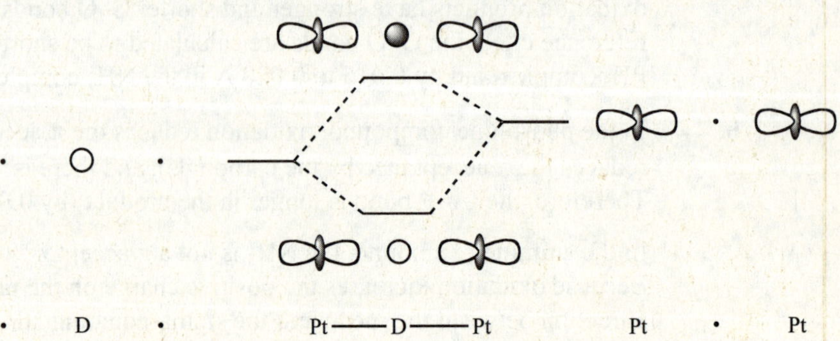

10.31 MnO_4^- has no *d* electrons, while MnO_4^{2-} has one. The slight antibonding effect of this electron is enough to lengthen the bonds. In addition, there is less electrostatic attraction by Mn(VI) in MnO_4^{2-} than by Mn(VII) in MnO_4^-.

10.32 **a.**

O_h	E	$8C_3$	$6C_2$	$6S_4$	$3C_2(=C_4^2)$	i	$6S_4$	$8S_6$	$3\sigma_h$	$6\sigma_d$	
Γ	6	0	0	2	2	0	0	0	4	2	
A_{1g}	1	1	1	1	1	1	1	1	1	1	
E_g	2	−1	0	0	2	2	0	−1	2	0	$(2z^2-x^2-y^2, x^2-y^2)$
T_{1u}	3	0	−1	1	−1	−3	−1	0	1	1	(x, y, z)

b. The representation reduces to $A_{1g} + E_g + T_{1u}$, as shown above.

c. The matching orbitals on Ti are: A_{1g}: s ; E_g: $d_{x^2-y^2}, d_{z^2}$; T_{1u}: p_x, p_y, p_z

d.

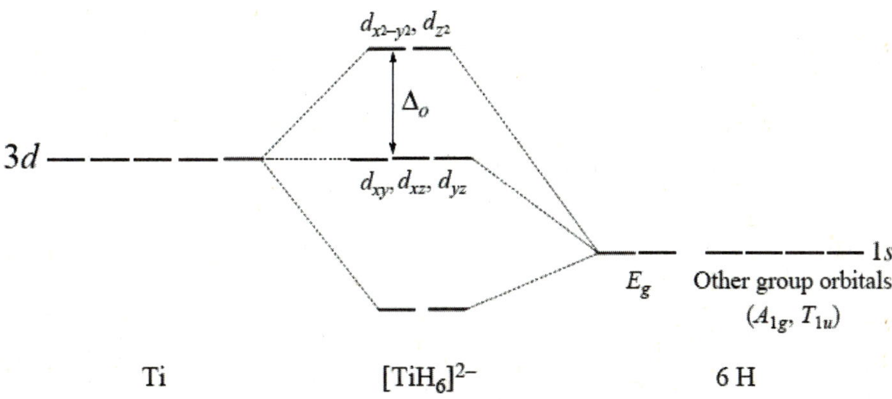

10.33 **a.** The t_{2g} molecular orbitals should involve titanium's d_{xy}, d_{xz}, and d_{yz} orbitals, the the e_g orbitals should involve the metal's $d_{x^2-y^2}$, d_{z^2} orbitals.

b. A difference from Figure 10.6 that should be noted is that the t_{2g} orbitals of the metal are no longer nonbonding in TiF_6^{3-} but are involved in the formation of bonding and antibonding molecular orbitals. In comparison with Figure 10.8, the t_{2g} orbitals with greatest metal character should form molecular orbitals that are higher rather than lower in energy in comparison with the d_{xy}, d_{xz}, and d_{yz} orbitals of the free metal ion.

10.34 **a.** The iron orbitals matching the representation *e* are the $d_{x^2-y^2}$ and d_{z^2} ; the orbitals matching t_2 are the d_{xy}, d_{xz}, and d_{yz}. The *d* orbital–ligand orbital interactions should appear more indirect than in the case of an octahedral transition metal complex because the lobes of these orbitals do not point directly toward the ligands.

b. The results for $FeCl_4^-$ should be simpler than shown in Figure 10.20 because the chloro ligand does not have π* orbitals to be taken into consideration (the LUMO orbitals of CO).

10.35 Results may vary considerably, depending on the software used. It is probably more likely that selecting the π-acceptor ligand (CN^-), a σ-donor, and a π-donor will show a trend matching that in Table 10.12 than if a set of three of σ -donor or π-donor ligands is used—but examining each type can be instructive.

CHAPTER 11: COORDINATION CHEMISTRY III: ELECTRONIC SPECTRA

11.1 **a.** p^3 There will be (6!)/(3!3!) = 20 microstates

		M_S			
		–3/2	–1/2	1/2	3/2
	+2		$1^+\ 1^-\ 0^-$	$1^+\ 1^-\ 0^+$	
	+1		$1^+\ 1^-\ -1^-$ $1^-\ 0^-\ 0^+$	$1^+\ 1^-\ -1^+$ $1^+\ 0^-\ 0^+$	
M_L	0	$1^-\ 0^-\ -1^-$	$1^+\ 0^-\ -1^-$ $1^-\ 0^+\ -1^-$ $1^-\ 0^-\ -1^+$	$1^+\ 0^+\ -1^-$ $1^+\ 0^-\ -1^+$ $1^-\ 0^+\ -1^+$	$1^+\ 0^+\ -1^+$
	–1		$-1^+\ -1^-\ 1^-$ $-1^-\ 0^-\ 0^+$	$-1^+\ -1^-\ 1^+$ $-1^+\ 0^-\ 0^+$	
	–2		$-1^+\ -1^-\ 0^-$	$-1^+\ -1^-\ 0^+$	

Terms: $L = 0, S = 3/2$: $^4\boldsymbol{S}$ (ground state)
$L = 2, S = 1/2$: $^2\boldsymbol{D}$
$L = 1, S = 1/2$: $^2\boldsymbol{P}$

b. p^1d^1 There will be (6!)/(1!5!) (10!)/(1!9!) = 60 microstates

		M_S		
		–1	0	1
	3	$1^-\ 2^-$	$1^-\ 2^+,\ \ 1^+\ 2^-$	$1^+\ 2^+$
	2	$1^-\ 1^-$ $0^-\ 2^-$	$1^+\ 1^-,\ \ 1^-\ 1^+$ $0^-\ 2^+,\ \ 0^+\ 2^-$	$1^+\ 1^+$ $0^+\ 2^+$
	1	$1^-\ 0^-$ $0^-\ 1^-$ $-1^-\ 2^-$	$1^+\ 0^-,\ \ 0^+\ 1^-$ $1^-\ 0^+,\ \ 0^-\ 1^+$ $-1^-\ 2^+,\ \ -1^+\ 2^-$	$1^+\ 0^+$ $0^+\ 1^+$ $-1^+\ 2^+$
M_L	0	$1^-\ -1^-$ $0^-\ 0^-$ $-1^-\ 1^-$	$1^+\ -1^-,\ \ 1^-\ -1^+$ $0^-\ 0^+,\ \ 0^+\ 0^-$ $-1^+\ 1^-,\ \ -1^-\ 1^+$	$1^+\ -1^+$ $0^+\ 0^+$ $-1^+\ 1^+$
	–1	$-1^-\ 0^-$ $0^-\ -1^-$ $1^-\ -2^-$	$-1^+\ 0^-,\ \ 0^+\ -1^-$ $-1^-\ 0^+,\ \ 0^-\ -1^+$ $1^-\ -2^+,\ \ 1^+\ -2^-$	$-1^+\ 0^+$ $0^+\ -1^+$ $1^+\ -2^+$
	–2	$-1^-\ -1^-$ $0^-\ -2^-$	$-1^+\ -1^-,\ -1^-\ -1^+$ $0^+\ -2^-,\ 0^-\ -2^+$	$-1^+\ -1^+$ $0^+\ -2^+$
	–3	$-1^-\ -2^-$	$-1^-\ -2^+,\ -1^+\ -2^-$	$-1^+\ -2^+$

Terms: $L = 3, S = 1$ $^3\boldsymbol{F}$ (ground state) $L = 2, S = 0$ $^1\boldsymbol{D}$
$L = 3, S = 0$ $^1\boldsymbol{F}$ $L = 1, S = 1$ $^3\boldsymbol{P}$
$L = 2, S = 1$ $^3\boldsymbol{D}$ $L = 1, S = 0$ $^1\boldsymbol{P}$

The two electrons have quantum numbers that are independent of each other, because the electrons are in different orbitals. Because they have different l values, the electrons can have the same m_l and m_s values.

11.2 For p^3: $L = 0$, $S = 3/2$, the term ${}^4S$ has $J = 3/2$ only ($|L+S| = |L–S|$). Therefore, the ground state is $\boldsymbol{{}^4S_{3/2}}$.

For p^1d^1: $L = 3$, $S = 1$, the term ${}^3F$ has $J = 4, 3, 2$. Since both levels are less than half filled, the state having lowest J has lowest energy, and the ground state is $\boldsymbol{{}^3F_2}$.

11.3 **a.** s^1d^1 There are 20 microstates:

M_L \ M_S	-1	0	$+1$
$+2$	$0^-\ 2^-$	$0^-\ 2^+,\ 0^+\ 2^-$	$0^+\ 2^+$
$+1$	$0^-\ 1^-$	$0^-\ 1^+,\ 0^+\ 1^-$	$0^+\ 1^+$
0	$0^-\ 0^-$	$0^-\ 0^+,\ 0^+\ 0^-$	$0^+\ 0^+$
-1	$0^-\ {-1}^-$	$0^-\ {-1}^+,\ 0^+\ {-1}^-$	$0^+\ {-1}^+$
-2	$0^-\ {-2}^-$	$0^-\ {-2}^+,\ 0^+\ {-2}^-$	$0^+\ {-2}^+$

b. Terms: $L = 2$, $S = 1$: $\boldsymbol{{}^3D}$; $L = 2$, $S = 0$: $\boldsymbol{{}^1D}$

c. The $\boldsymbol{{}^3D}$, with the highest spin multiplicity, is the lowest energy term.

11.4 **a.** d^1f^1

M_L \ M_S	-1	0	$+1$
$+5$	$2^-\ 3^-$	$2^-\ 3^+,\ 2^+\ 3^-$	$2^+\ 3^+$
$+4$	$2^-\ 2^-$ $1^-\ 3^-$	$2^-\ 2^+,\ 2^+\ 2^-$ $1^-\ 3^+,\ 1^+\ 3^-$	$2^+\ 2^+$ $1^+\ 3^+$
$+3$	$2^-\ 1^-$ $1^-\ 2^-$ $0^-\ 3^-$	$2^-\ 1^+,\ 2^+\ 1^-$ $1^-\ 2^+,\ 1^+\ 2^-$ $0^-\ 3^+,\ 0^+\ 3^-$	$2^+\ 1^+$ $1^+\ 2^+$ $0^+\ 3^+$
$+2$	$2^-\ 0^-$ $1^-\ 1^-$ $0^-\ 2^-$ ${-1}^-\ 3^-$	$2^-\ 0^+,\ 2^+\ 0^-$ $1^-\ 1^+,\ 1^+\ 1^-$ $0^-\ 2^+,\ 0^+\ 2^-$ ${-1}^-\ 3^+,\ {-1}^+\ 3^-$	$2^+\ 0^+$ $1^+\ 1^+$ $0^+\ 2^+$ ${-1}^+\ 3^+$
$+1$	$2^-\ {-1}^-$ $1^-\ 0^-$ $0^-\ 1^-$ ${-1}^-\ 2^-$ ${-2}^-\ 3^-$	$2^-\ {-1}^+,\ 2^+\ {-1}^-$ $1^-\ 0^+,\ 1^+\ 0^-$ $0^-\ 1^+,\ 0^+\ 1^-$ ${-1}^-\ 2^+,\ {-1}^+\ 2^-$ ${-2}^-\ 3^+,\ {-2}^-\ 3^-$	$2^+\ {-1}^+$ $1^+\ 0^+$ $0^+\ 1^+$ ${-1}^+\ 2^+$ ${-2}^-\ 3^+$
0	$2^-\ {-2}^-$ $1^-\ {-1}^-$ $0^-\ 0^-$ ${-1}^-\ 1^-$ ${-2}^-\ 2^-$	$2^-\ {-2}^+,\ 2^+\ {-2}^-$ $1^-\ {-1}^+,\ 1^+\ {-1}^-$ $0^-\ 0^+,\ 0^+\ 0^-$ ${-1}^-\ 1^+,\ {-1}^+\ 1^-$ ${-2}^-\ 2^+,\ {-2}^+\ 2^-$	$2^+\ {-2}^+$ $1^+\ {-1}^+$ $0^+\ 0^+$ ${-1}^+\ 1^+$ ${-2}^+\ 2^+$
-1	${-2}^-\ 1^-$ ${-1}^-\ 0^-$ $0^-\ {-1}^-$ $1^-\ {-2}^-$ $2^-\ {-3}^-$	${-2}^-\ 1^+,\ {-2}^+\ 1^-$ ${-1}^-\ 0^+,\ {-1}^+\ 0^-$ $0^-\ {-1}^+,\ 0^+\ {-1}^-$ $1^-\ {-2}^+,\ 1^+\ {-2}^-$ $2^-\ {-3}^+,\ 2^-\ {-3}^-$	${-2}^+\ 1^+$ ${-1}^+\ 0^+$ $0^+\ {-1}^+$ $1^+\ {-2}^+$ $2^-\ {-3}^+$

continued

–2	$-2^- \; 0^-$ $-1^- -1^-$ $0^- -2^-$ $1^- -3^-$	$-2^- \; 0^+$, $-2^+ \; 0^-$ $-1^- -1^+$, $-1^+ -1^-$ $0^- -2^+$, $0^+ -2^-$ $1^- -3^+$, $1^+ -3^-$	$-2^+ \; 0^+$ $-1^+ -1^+$ $0^+ -2^+$ $1^+ -3^+$
–3	$-2^- -1^-$ $-1^- -2^-$ $0^- -3^-$	$-2^- -1^+$, $-2^+ -1^-$ $-1^- -2^+$, $-1^+ -2^-$ $0^- -3^+$, $0^+ -3^-$	$-2^+ -1^+$ $-1^+ -2^+$ $0^+ -3^+$
–4	$-2^- -2^-$ $-1^- -3^-$	$-2^- -2^+$, $-2^+ -2^-$ $-1^- -3^+$, $-1^+ -3^-$	$-2^+ -2^+$ $-1^+ -3^+$
–5	$-2^- -3^-$	$-2^- -3^+$, $-2^+ -3^-$	$-2^+ -3^+$

b. Terms: $L = 5, S = 1$: $^3\boldsymbol{H}$; $L = 5, S = 0$: $^1\boldsymbol{H}$; $L = 4, S = 1$: $^3\boldsymbol{G}$; $L = 4, S = 0$: $^1\boldsymbol{G}$; $L = 3, S = 1$: $^3\boldsymbol{F}$; $L = 3, S = 0$: $^1\boldsymbol{F}$; $L = 2, S = 1$: $^3\boldsymbol{D}$; $L = 2, S = 0$: $^1\boldsymbol{D}$; $L = 1, S = 1$: $^3\boldsymbol{P}$; $L = 1, S = 0$: $^1\boldsymbol{P}$

c. The lowest energy term is the $^3\boldsymbol{H}$. For this term, J has the values 6, 5, and 4. Because the subshells are less than half full, the lowest value of J provides the lowest energy: $^3\boldsymbol{H}_4$.

11.5 a. From problem 11.1a, for a p^3 configuration there are three terms, ^4S, ^2D, and ^2P. The J values for each of these are determined below.

For ^4S: $L = 0, S = 3/2$.

Because $J = L + S, L + S - 1, \ldots|L - S|$, the quantum number J can only be 3/2 and there is a single state for ^4S: $^4S_{3/2}$.

For ^2D: $L = 2, S = 1/2$.

Possible J values are 5/2 and 3/2, and the two possible states are $^2D_{5/2}$ and $^2D_{3/2}$.

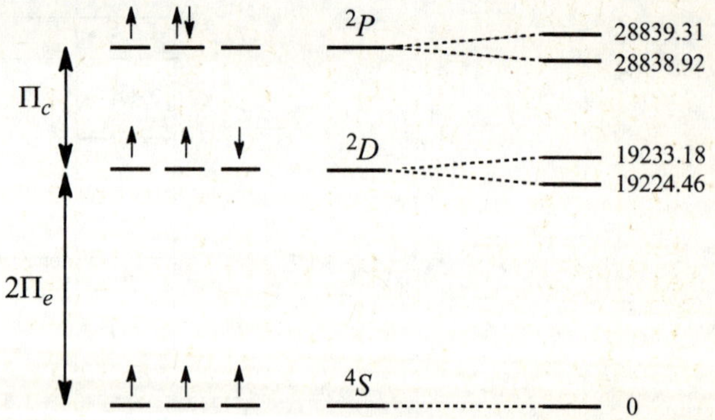

^2P 28839.31 28838.92
Π_c
^2D 19233.18 19224.46
$2\Pi_e$
^4S 0

For ^2P: $L = 1, S = 1/2$. Possible J values are 3/2 and 1/2, with states $^2P_{5/2}$ and $^2P_{1/2}$.

The lowest energy state is $^4S_{3/2}$ (highest multiplicity). $^2D_{5/2}$ and $^2D_{3/2}$ are next, at 19233.18 and 19224.46 cm^{-1}, and $^2P_{3/2}$ and $^2P_{1/2}$ are the highest energy at 28838.92 and 28839.31 cm^{-1}.

b. The difference in energy between the ^4S and ^2D states is $2\Pi_e$. From the average of the two nearly degenerate ^2D states, $\Pi_e = 9614.41$ cm^{-1}.

The difference in energy between the averages of the ^2D and ^2P states is $\Pi_c = 28{,}839.12 - 19{,}228.82 = 9610.30$ cm^{-1}.

11.6 **a.** ^2D has $L = 2$ and $S = 1/2$, so $M_L = -2, -1, 0, 1, 2$ and $M_S = -1/2, 1/2$

b. ^3G has $L = 4$ and $S = 1$, so $M_L = -4, -3, -2, -1, 0, 1, 2, 3, 4$ and $M_S = -1, 0, 1$

c. ^4F has $L = 3$ and $S = 3/2$, so $M_L = -3, -2, -1, 0, 1, 2, 3$ and $M_S = -3/2, -1/2, 1/2, 3/2$

11.7 **a.** ^2D with $J = 5/2, 3/2$ fits an excited state of d^3, $^2D_{3/2}$

b. ^3G with $J = 5, 4, 3$ fits an excited state of d^4, $^3G_3$

c. ^4F with $J = 9/2, 7/2, 5/2, 3/2$ fits the ground state of d^7, $^4F_{9/2}$

11.8

$$\varepsilon = 0.038\frac{\text{L}}{\text{mol cm}} \quad A = 0.10 \quad l = 1.00\text{cm}$$

$$A = \varepsilon lc;\ c = \frac{A}{\varepsilon l} = \frac{0.10}{\left(0.038\frac{\text{L}}{\text{mol cm}}\right)(1.00\text{cm})} = 2.6\frac{\text{mol}}{\text{L}}$$

11.9 **a.**

$$\bar{\nu} = 24{,}900\ \text{cm}^{-1}$$

$$\lambda = \frac{1}{\bar{\nu}} = \frac{1}{24{,}900\ \text{cm}^{-1}} = 4.02\times10^{-5}\text{cm} = 402\ \text{nm}$$

$$\nu = \frac{c}{\lambda} = \frac{(2.998\times10^8\text{m s}^{-1})(100\ \text{cm m}^{-1})}{4.02\times10^{-5}\text{cm}} = 7.47\times10^{14}\text{s}^{-1}$$

b.

$$\lambda = 366\ \text{nm}$$

$$\nu = \frac{c}{\lambda} = \frac{(2.998\times10^8\text{m s}^{-1})(10^9\text{nm m}^{-1})}{366\ \text{nm}} = 8.19\times10^{14}\text{s}^{-1}$$

$$E = h\nu = (6.626\times10^{-34}\ \text{Js})(8.19\times10^{14}\text{s}^{-1}) = 5.43\times10^{-19}\text{J}$$

11.10 J values are included in these answers:

a. d^8 O_h $M_S = 1 = S$ Spin multiplicity = 2 + 1 = 3
Max M_L = 2+2+1+1+0+0–1–2 = 3 = L, so F term. $J = 4,3,2$ $\mathbf{^3F_4}$

b. d^5 O_h high spin $M_S = 5/2 = S$ Spin multiplicity = 5 + 1 = 6
Max M_L = 2+1+0–1–2 = 0 = L, so S term. $J = 5/2$ $\mathbf{^6S_{5/2}}$

d^5 O_h low spin $M_S = 1/2 = S$ Spin multiplicity = 1 + 1 = 2
Max M_L = 2+2+1+1+0 = 6 = L, so I term. $J = |L \pm S| = 11/2, 13/2$ $\mathbf{^2I_{11/2}}$, $\mathbf{^2I_{13/2}}$
(J is uncertain in this case; the usual rule does not apply because the level is exactly half full)

c. d^4 T_d $M_S = 2 = S$ Spin multiplicity = 4+1=5
Max M_L = 2+1+0–1 = 2 = L, so D term. $J = 4, 3, 2, 1, 0$ $\mathbf{^5D_0}$

d. d^9 D_{4h} $M_S = 1/2 = S$ Spin multiplicity = 1+1=2
Max M_L = 2+2+1+1+0+0–1–1–2 = 2 = L, so D term. $J = |L \pm S| = 5/2, 3/2$ $\mathbf{^2D_{5/2}}$

11.11 $[Ni(H_2O)_6]^{2+}$ For d^8 ions, the energy of the lowest energy band is Δ_o, so $\Delta_o = 8{,}700\ cm^{-1}$. The bands are split due to Jahn-Teller distortion in the excited state.

11.12 **a.** $[Cr(C_2O_4)_3]^{3-}$ is Cr(III), d^3. Δ_o is equal to the lowest energy band, so $\Delta_o = 17{,}400\ cm^{-1}$.

b. $[Ti(NCS)_6]^{3-}$ is Ti(III), d^1. Δ_o is the energy of the single band; $\Delta_o = 18{,}400\ cm^{-1}$. The band is split due to Jahn-Teller distortion of the excited state.

c. $[Ni(en)_3]^{2+}$ is Ni(II), d^8. The lowest energy band corresponds to Δ_o; $\Delta_o = 11{,}200\ cm^{-1}$.

d. $[VF_6]^{3-}$ is V(III), d^2. Following the example on pp. 434–435, we find the ratio ν_2/ν_1 and then Δ_o /B: $\nu_2/\nu_1 = 1.57$ at Δ_o /B = 26. From the Tanabe-Sugano diagram at $\Delta_o = 26$,

ν_1: E/B = 24.1   E = 24.1 B = 14,800 cm^{-1} B = 614 cm^{-1}

ν_2: E/B = 37.0   E = 37.0 B = 23,250 cm^{-1} B = 628 cm^{-1}

Average B = 620 cm^{-1}, Δ_o = 26 B = 16,000 cm^{-1}

e. V(III) is a d^2 ion. Again following the example on pp. 434–435, ν_2 = 21,413 cm^{-1} and ν_1 = 14,409 cm^{-1}, $\nu_2/\nu_1 = 1.49$. From the Tanabe-Sugano diagram at $\Delta_o = 34.5$,

ν_1: E/B = 29   E = 29 B =14,409 cm^{-1} B = 489 cm^{-1}

ν_2: E/B = 44   E = 44 B = 21,413 cm^{-1} B = 489 cm^{-1}

Average B = 489 cm^{-1}, Δ_o = 34.5 B = 16,870 cm^{-1}

11.13 $[Co(NH_3)_6]^{2+}$, d^7. As in problem 11.12d:
$\nu_2/\nu_1 = 2.34$ at Δ_o/B = 11. From the Tanabe-Sugano diagram at $\Delta_o = 11$,

ν_1: E/B = 10   E= 10 B = 9,000 cm^{-1} B = 900 cm^{-1}

ν_2: E/B = 22.5   E= 22.5 B = 21,100 cm^{-1} B = 938 cm^{-1}

Average B = 920 cm^{-1}, Δ_o = 11 B = 10,100 cm^{-1}

11.14 **a.** $t_{2g}^4 e_g^2$ The t_{2g} level is a triply degenerate asymmetrically occupied state, so it is *T*.

b. t_{2g}^6 This is a nondegenerate state, completely occupied, so it is *A*.

c. $t_{2g}^3 e_g^3$ This is an excited state, with the t_{2g} level uniformly occupied and the e_g level doubly degenerate, so it is *E*.

d. t_{2g}^5 This is a triply degenerate state, *T*. (Vacancies in the orbitals can be treated similarly to electrons.)

e. e_g Another excited state, this is doubly degenerate, *E*.

11.15 The complexes with potential degeneracies are those with d^1, d^2, d^4, d^7, and d^9, low spin d^5, and high spin d^6 configurations. The strongest effects are with high spin d^4, low spin d^7, and d^9 complexes, corresponding to $[Mn(NH_3)_6]^{3+}$, $[Ni(NH_3)_6]^{3+}$, and $[Cu(NH_3)_6]^{3+}$. Weaker effects might be seen with $[Ti(NH_3)_6]^{3+}$, $[V(NH_3)_6]^{3+}$, low spin $[Fe(NH_3)_6]^{3+}$, and high spin $[Co(NH_3)_6]^{3+}$, all of which are unknown or unstable. Other ligands are needed to stabilize ions containing these 3+ metal ions.

11.16 Rhenium is significantly larger than manganese, and the $5d$ orbitals of Re are higher in energy than the $3d$ orbitals of Mn, so a CTTM excitation requires more energy for ReO_4^-. In addition, since the molecular orbitals derived primarily from the $3d$ orbitals of MnO_4^- are lower in energy than the corresponding MO's of ReO_4^-, MnO_4^- is better able to accept electrons; it is a better oxidizing agent.

11.17 The order of energy of the charge transfer bands is I < Br < Cl. The HOMO of the iodide complex is highest in energy, and the LUMO's are similar in energy, so the difference is smallest for iodide. In other terms, iodide is softer and can lose an electron most easily in a CTTM process.

11.18 Comparing $[Fe(CN)_6]^{3-}$ (low spin d^5) and $[Fe(CN)_6]^{4-}$ (low spin d^6), where CN^- is a σ donor and a π acceptor: the t_{2g} orbitals of $[Fe(CN)_6]^{3-}$ contain 5 electrons, allowing CTTM from the ligand orbitals to either t_{2g} or e_g levels. The t_{2g} levels of $[Fe(CN)_6]^{4-}$ are full, so only the higher e_g levels are available for CTTM. CTTL transitions ($t_{2g} \rightarrow \pi^*$ excitations) are also possible for either complex.

11.19 **a.** O^{2-} and Cl^- are both σ and π donors, and the metal ions are Cr(V) and Mo(V) (the ligands are Cl^- and O^{2-}). Metal d orbitals, influence of Cl^- ligands:

	e_σ	e_π	Total
z^2	2	0	$2\ e_\sigma$
x^2-y^2	3	2	$3\ e_\sigma + 2\ e_\pi$
xy	0	2	$2\ e_\pi$
xz	0	2	$2\ e_\pi$
yz	0	2	$2\ e_\pi$

Metal d orbitals, influence of O^{2-} ligands:

	e_σ	e_π	Total
z^2	1	0	e_σ
x^2-y^2	0	0	0
xy	0	0	0
xz	0	1	e_π
yz	0	1	e_π

Overall, from bottom to top, the energies of d orbitals based on angular overlap are:

xy	$2e_\pi$(Cl)
xz and yz	$2e_\pi$(Cl) + e_π(O)
z^2	$2e_\sigma$(Cl) + e_σ(O)
x^2-y^2	$3e_\sigma$(Cl) + $2e_\pi$(Cl)

b. These complexes have C_{4v} symmetry. The symmetry labels of the orbitals are given in the character table on the next page.

C_{4v}	E	$2C_4$	C_2	$2\sigma_v$	$2\sigma_d$	
A_1	1	1	1	1	1	z^2
B_1	1	−1	1	1	−1	x^2-y^2
B_2	1	−1	1	−1	1	xy
E	2	0	−2	0	0	(xz, yz)

Each of the representations is labeled with the matching orbital(s).

c. The lowest d orbital has an energy of $2e_\pi$(Cl); the next have energies of $2e_\pi$(Cl) + e_π(O). Because these are d^1 complexes, the transition is between these orbitals; it is from an essentially nonbonding orbital to a π* orbital. The interactions between metal and ligand are generally stronger for a second row transition metal than for the first row, raising the π* energy in the Mo case.

11.20 $[V(CO)_6]^- < Cr(CO)_6 < [Mn(CO)_6]^+$ As the nuclear charge on the metal increases, the metal orbitals are drawn to lower energies. Consequently, the CTTL bands should increase in energy.

11.21 **a.** At 80K: $\mu_S = 0.65 = \sqrt{n(n+2)}$ $n = 0.19$

At 300K: $\mu_S = 5.2 = \sqrt{n(n+2)}$ $n = 4.3$

b. The complex is near the low spin-high spin boundary of the d^6 Tanabe-Sugano diagram. High spin becomes increasingly favored as the temperature increases.

11.22 **a.** ML_2, using positions 1 and 6, with O^{2-} both a σ and π donor:

	e_σ	e_π	Total
z^2	2	0	$2\ e_\sigma$
x^2-y^2	0	0	0
xy	0	0	0
xz	0	2	$2\ e_\pi$
yz	0	2	$2\ e_\pi$

b. If this is a high spin complex, there are four electrons in the lowest levels (xy, x^2-y^2), three in the next two (xz, yz), and one in the highest (z^2). Electronic transitions can be from the middle levels to the top, and from the bottom levels to the middle and the top—three possibilities in all. According to the reference, the transitions seen are from the middle and the bottom levels to the top level.

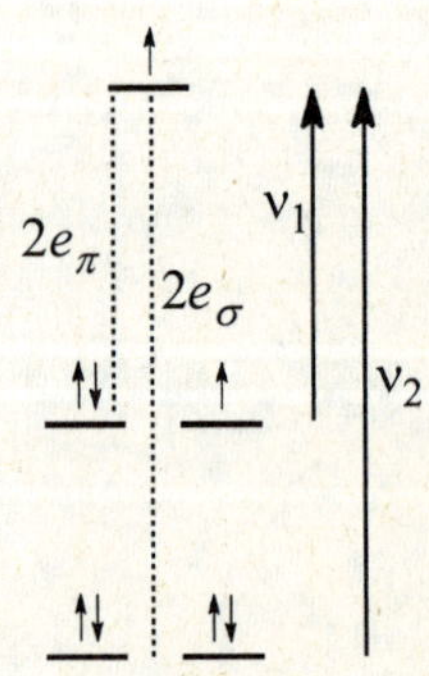

c. Assigning the transitions as in part b:

$E = 2e_\sigma = 16{,}000\ \text{cm}^{-1}$ $E = 2e_\pi = 9{,}000\ \text{cm}^{-1}$

so $e_\sigma = 8{,}000\ \text{cm}^{-1}$ and $e_\pi = 4{,}500\ \text{cm}^{-1}$

The reference provides a much more detailed analysis, including a discussion of the paramagnetism of this complex and other factors related to its electronic spectrum.

11.23 $Re(CO)_3(P(OPh_3))(DBSQ)$ 18,250 cm^{-1}
$Re(CO)_3(PPh_3)(DBSQ)$ 17,300 cm^{-1}
$Re(CO)_3(NEt_3)(DBSQ)$ 16,670 cm^{-1}

NEt_3 is the strongest donor ligand in this series. Therefore, the metal in the complex $Re(CO)_3(NEt_3)(DBSQ)$ has the greatest concentration of electrons and the greatest tendency to transfer electrons to acceptor orbitals. Since this complex also has the lowest energy charge transfer band, we may assign this as charge transfer to ligand.

11.24 **a.** RuO_4^{2-} has the highest value of Δ_t. Δ_t increases with the oxidation state of the metal and in general is greater for second row than for first row metals. The overall trend is $RuO_4^{2-} > FeO_4^{2-} > MnO_4^{3-} > CrO_4^{4-}$.

b. The nuclear charge of iron is greatest in this isoelectronic series and exerts the strongest attraction for bonding electrons. As a result, FeO_4^{2-} has the shortest metal–oxygen distance, 165 pm, in comparison with 170 pm for MnO_4^{3-} and 176 pm for CrO_4^{4-} and RuO_4^{2-}.

c. As the nuclear charge of the metal increases, the metal orbitals are pulled in to lower energies. Consequently, less energy is needed to excite electrons from ligand orbitals to metal orbitals. These are ligand to metal charge transfers.

11.25 Addition of aqueous NH_3 replaces the H_2O ligands in the green $[Ni(H_2O)_6]^{2-}$ to give blue $[Ni(NH_3)_6]^{2+}$. As a bidentate ligand, en can replace two NH_3 ligands; three en ligands can replace all six NH_3 ligands to form violet $[Ni(en)_3]^{2+}$:

	$[Ni(H_2O)_6]^{2-}$	$\xrightarrow{NH_3}$	$[Ni(NH_3)_6]^{2+}$	$\xrightarrow{en}$	$[Ni(en)_3]^{2+}$
observed color:	green		blue		violet
complementary color:	red		orange		yellow

The complementary colors in this series have increasing energies, indicating that en has the strongest effect on Δ_o, H_2O the weakest. This is consistent with the positions of these ligands in the spectrochemical series.

11.26 **a.** These colors are most likely the consequence of CTTM transitions, from orbitals that are primarily from the oxide ligands to orbitals that are primarily from the metal:

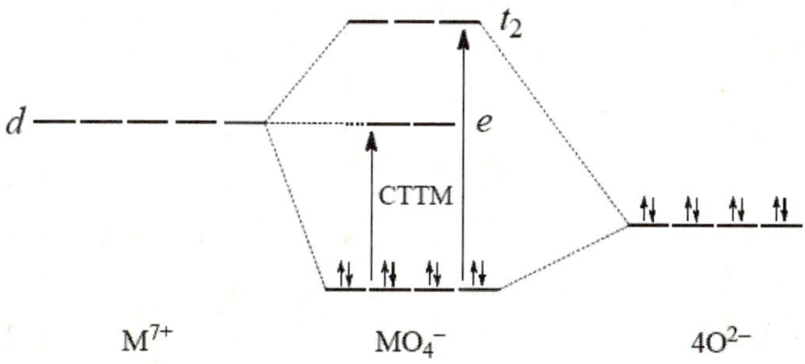

b. In TcO_4^-, the separation between the donor orbitals of the O^{2-} ligands and the acceptor orbitals is greater than in MnO_4^-. As a consequence, TcO_4^- absorbs light of higher energy (green) than MnO_4^- (yellow). Actually, most of the absorption by TcO_4^- is in the ultraviolet, with the pale red color a result of a tail of the absorption band extending into the visible.

c. The metal-ligand interactions in MnO_4^{2-} (Mn(VI)) are weaker than in MnO_4^- (Mn(VII)), and the separation of donor and acceptor orbitals in MnO_4^{2-} is smaller, meaning that less energy (red light) is necessary for excitation than in MnO_4^- (yellow). Also worth noting: the Mn—O bond distance is longer in MnO_4^{2-} (165.9 pm) than in MnO_4^- (162.9 pm), an indication of weaker bonding in the former.

11.27 a. At 350 nm:

$$\varepsilon = \frac{A}{lc} = \frac{2.34}{(1.00\text{cm})(2.00\times10^{-4}\text{ M})} = 11{,}700\text{ L mol}^{-1}\text{ cm}^{-1}$$

At 514 nm:

$$\varepsilon = \frac{A}{lc} = \frac{0.532}{(1.00\text{cm})(2.00\times10^{-4}\text{ M})} = 2{,}640\text{ L mol}^{-1}\text{ cm}^{-1}$$

At 590 nm:

$$\varepsilon = \frac{A}{lc} = \frac{0.370}{(1.00\text{cm})(2.00\times10^{-4}\text{ M})} = 1{,}850\text{ L mol}^{-1}\text{ cm}^{-1}$$

At 1540 nm:

$$\varepsilon = \frac{A}{lc} = \frac{0.0016}{(1.00\text{cm})(2.00\times10^{-4}\text{ M})} = 8.0\text{ L mol}^{-1}\text{ cm}^{-1}$$

b. Because of their high intensity, the bands at 350, 514, and 590 nm are probably charge transfer bands. However, the low molar absorptivity of the band at 1540 nm indicates that it is probably a *d–d* transition (see examples in Figure 11.8).

11.28 These are all d^8 complexes, with three excited states of the same spin multiplicity as the ground state. For d^8, Δ_o = energy of the lowest energy band. *B* can be calculated by using the method of Problem 11.12. Δ_o is the difference between the transitions $^3A_2 \rightarrow {}^3T_1$ and $^3A_2 \rightarrow {}^3T_2$; a graph of ν_2/ν_1 versus Δ_o/B needs to be prepared for d^8 Ni^{2+} in order to calculate *B*, as described in pages 434–435. A plot of the ratio of the highest energy band to the lowest energy band is shown here.

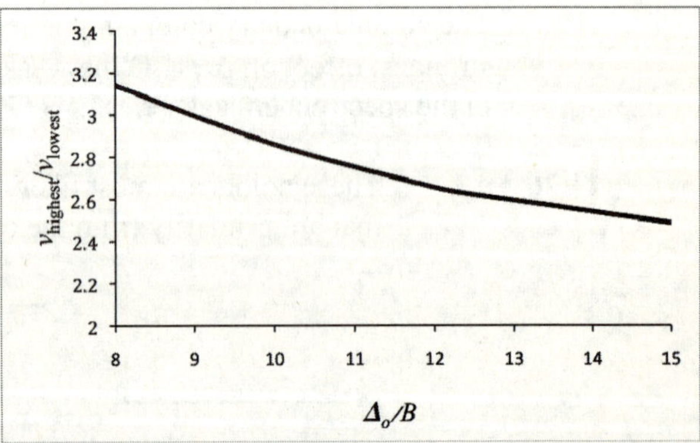

Species	Ratio	Δ_o/B	Δ_o(cm^{-1})	*B*(cm^{-1})
$[Ni(H_2O)_6]^{2+}$	3.06	8.8	8,500	970
$[Ni(NH_3)_6]^{2+}$	2.62	12.5	10,500	860
$[Ni(OS(CH_3)_2)_6]^{2+}$	3.11	8.4	7,728	920
$[Ni(dma)_6]^{2+}$	3.14	8.0	7,576	950

11.29 **a.** These are both high spin d^7 complexes, for which the ground state term symbol is ^4F (see Figure 11.7).

b. There are three possible transitions, all originating from the $^4T_1(^4F)$ and going to the $^4T_2$, $^4A_2$, and $^4T_1(^4P)$ levels.

c. For $[Co(bipy)_3]^{2+}$, the ratio $\frac{\nu_2}{\nu_1} = \frac{22{,}000 \text{ cm}^{-1}}{11{,}300 \text{ cm}^{-1}} = 1.95$. From Figure 11.14, this comes at $\frac{\Delta_o}{B} = 17$, at which:

for the 22000 cm^{-1} band: $\frac{E}{B} = 28$ and $B = \frac{22{,}000 \text{ cm}^{-1}}{28} = 786 \text{ cm}^{-1}$

for the 11300 cm^{-1} band: $\frac{E}{B} = 15$ and $B = \frac{11{,}300 \text{ cm}^{-1}}{15} = 754 \text{ cm}^{-1}$

Average $B = 770 \text{ cm}^{-1}$; $\Delta_o = 17 \quad B = 13{,}100 \text{ cm}^{-1}$

For a d^7 complex, LFSE = $-\frac{4}{5}\Delta_o = -10{,}500 \text{ cm}^{-1}$

Δ_o for $[Co(NH_3)_6]^{2+}$ was calculated in Problem 11.13; $\Delta_o = 10{,}100 \text{ cm}^{-1}$

For this d^7 complex, LFSE = $-\frac{4}{5}\Delta_o = -8{,}080 \text{ cm}^{-1}$

d. These bands should be broad (see Figure 11.8), a consequence of vibrational motion.

e. The molecular orbital energy level diagrams should be similar to Figure 10.5, with 6 electrons in t_{2g} orbitals and one electron in an e_g level in each case. The separation between t_{2g} and e_g orbitals should be larger in the bipy complex.

11.30 These absorption bands correspond to LMCT transitions. As the nuclear charge increases ($Fe \longrightarrow Ni$), the acceptor (largely *d*) orbitals decrease in energy, enabling the charge transfer transitions to occur at lower energy in the nickel complex.

Chapter 12: Coordination Chemistry IV: Reactions and Mechanisms

12.1 $[Cr(H_2O)_6]^{2+}$ is labile, and has 4 unpaired electrons, with one in the e_g level. This electron in an antibonding orbital makes substitution relatively easy. $[Cr(CN)_6]^{4-}$ is inert. It has all 4 electrons in the t_{2g} levels. Substitution requires activation into the antibonding e_g level, so reactions are slow. The strong π bonding with CN^- also strengthens the Cr–C bonds.

12.2 The kinetic equations for I_a and I_d mechanisms are similar, so using the rate equation to distinguish between them is not possible. However, if the rate constants for different entering ligands are quite different, it is clear that increasing the coordination number is important in the rate determining step. If dissociation is more important, the rates should be different for different leaving groups.

12.3 Pentachlorooxochromate(V) is a d^1 complex; it should be labile, with vacancies in the t_{2g} levels.

Hexaiodomanganate(IV) is a d^3 complex; it should be inert, much like Cr(III).

Hexacyanoferrate(III) is a low spin d^5 complex; it should be inert (vacant e_g levels).

Hexammineiron(III) is a d^6 high spin complex; with partly occupied orbitals in both levels, it should be labile.

12.4 The $[Fe(CN)_6]^{4-}$ ion is a low spin d^6 complex, with a maximum LFSE of $-2.4\ \Delta_o$. It is also a notably kinetically inert complex, hence its low reactivity even though the potentially toxic cyanide ligand is present.

12.5 $[Fe(H_2O)_6]^{3+}$ and $[Co(H_2O)_6]^{2+}$ are high-spin species, and the elctrons in the upper e_g levels make them labile. $[Cr(CN)_6]^{4-}$ is a d^4 low spin species. The t_{2g} levels are unequally occupied and the e_g are vacant, which makes it a borderline complex in terms of rate. $[Cr(CN)_6]^{3-}$, $[Fe(CN)_6]^{4-}$, and $[Co(NH_3)_5(H_2O)]^{3+}$ are all low-spin species with the t_{2g} levels either half filled or completely filled. This, combined with empty e_g levels, indicates inert species. LFAE numbers indicate that activation of these ions requires large amounts of energy.

12.6 The general rate law for square planar substitution is: Rate = $(k_1 + k_2[Cl^-])\ [Pt(NH_3)_4]^{2+}$.

The general procedure is to measure either the disappearance of $[Pt(NH_3)_4]^{2+}$ or the formation of $[Pt(NH_3)_3Cl]^+$ to find the rate. The most convenient method is by UV or visible absorbance spectra, using a wavelength where there is a large difference in absorbance between the two species.

a. Swamp the solution with excess chloride, so that its concentration does not change significantly. This is likely to make the reaction appear first order in Pt reactant. Measure the change in concentrations, plot $\ln[Pt(NH_3)_4]^{2+}$ vs. time to get k_{obs}.

b. Repeat several times with different chloride concentrations.

c. Graph k_{obs} vs. $[Cl^-]$. If the rate law above is followed, the result is a straight line with intercept = k_1 and slope = k_2.

If the rate equation above does not hold, the first order approximation in **a** will not work, and the logarithmic plot will not give a straight line.

12.7 **a.** Exchange with DMSO is extremely fast, regardless of the X^- concentration, because DMSO is present in very large excess as the solvent. The rate constant is also larger than for the other reactants.

b. The general equation for this reaction is probably: Rate = k [complex] [X]/(1+k'[X]). At low [X], the reactions are first order in [X] (1 >> k'[X]). At high [X], where 1 << k'[X], the reactions become zero order in [X].

c. The limiting rate constant is equal to k_2k_1 for D, and k_2K_1 for I_d.

d. Since the limiting rate constants are so similar, the major rate determining step is dissociative. However, the slight difference with differing X means that association has some part in the rate, and the mechanism is then I_d.

12.8 **a.** Because the rate is independent of the concentration of ^{13}CO, the rate determining step is most likely:

$$Cr(^{12}CO)_6 \longrightarrow Cr(^{12}CO)_5 + {}^{12}CO$$

The highly reactive $Cr(^{12}CO)_5$ then reacts rapidly with ^{13}CO.

b. These terms describe two pathways to product, a dissociative pathway (as in part **a**) and an associative pathway:

Dissociative:	$Cr(CO)_6 \longrightarrow Cr(CO)_5 + CO$	(slow)	rate = $k_1[Cr(CO)_6]$
	$Cr(CO)_5 + PR_3 \longrightarrow Cr(CO)_5(PR_3)$	(fast)	
Associative:	$Cr(CO)_6 + PR_3 \longrightarrow Cr(CO)_5(PR_3) + CO$		rate = $k_2[Cr(CO)_6][PR_3]$

With two mechanistic pathways leading to the same product, the overall rate is the sum of the rates of both.

c. Bulky ligands will tend to favor the first order (dissociative pathway) because the crowding around the metal will favor dissociation and hinder association with incoming ligands. (This effect is discussed further in section 14.1.1)

12.9 $[Cu(H_2O)_6]^{2+}$ is a d^9 complex, subject to Jahn-Teller distortion. Therefore, different rates are observed for exchange of axial and equatorial water molecules.

12.10 A plot of $[As(C_6H_5)_3]$ vs k_{obs} is a straight line, with intercept $k_1 = 2.3 \times 10^{-5}\ s^{-1}$ and slope $k_2 = 2.06 \times 10^{-5}\ M^{-1}s^{-1}$.

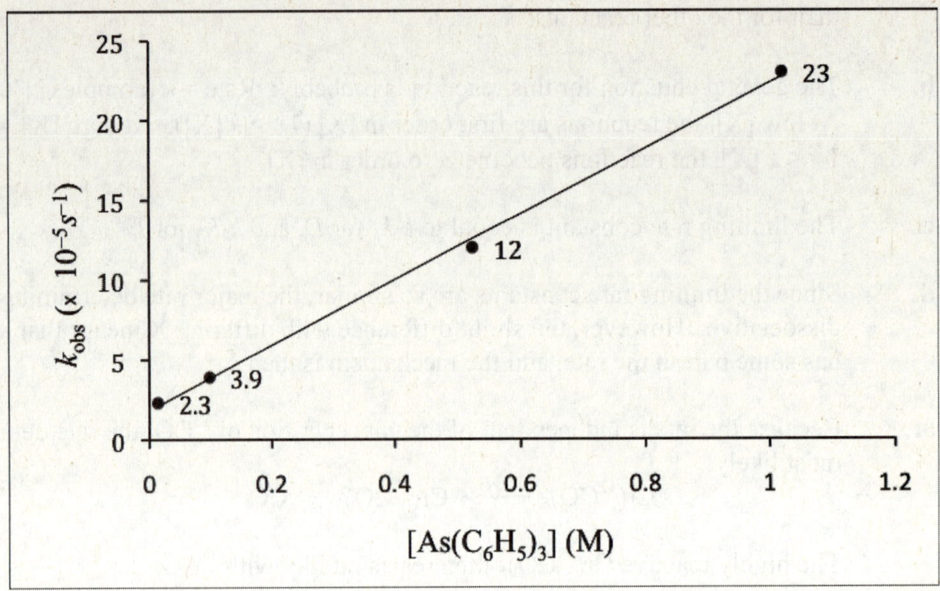

Rate = $(k_1 + k_2[As(C_6H_5)_3])[Co(NO)(CO)_3]$

This reaction, like the one in problem 6, shows both first and second order kinetics. The first order reaction appears to be a dissociative reaction or a solvent-assisted dissociation of CO, followed by a fast addition of $As(C_6H_5)_3$. The other path shows first order dependence on $As(C_6H_5)_3$, probably caused by an associative reaction.

12.11 ΔHNP is a measure of basicity, with more basic molecules having a smaller ΔHNP. The reactions appear to be associative, with the rate increasing with the basicity of the incoming ligand. The two lines are a consequence of the different natures of the ligands; numbers 1–11 are all phosphorus ligands, while 12–14 are organic nitrogen compounds. The half neutralization potential provides a relative measure of basicity within compounds of similar structure, but is not an absolute measure of the reactivity of the compounds.

12.12 **a.** $\Delta G° = \Delta H° - T\Delta S° = 10300 - 298 \times 55.6 = -6300$ J/mol = –6.3 kJ/mol

$\Delta G° = -RT \ln K$; $\ln K = -\Delta G°/RT = 6300/(8.3145 \times 298.15) = 2.53$; $K = 12.5$

b. The *cis* isomer has the higher bond energy (actually, the lower overall energy), since rearrangement to the *trans* isomer requires addition of energy. Since the phosphines are better π acceptors, the *cis* isomer should be the more stable. If the phosphines are *trans* to each other, they compete for the electron density of the same *d* orbital electrons. In the *cis* position, each can use one of the pair of d_{xz} and d_{yz} orbitals and avoid competition.

c. The free phosphine must help the reaction by an associative mechanism. Since benzene is the solvent, and it is very nonpolar, it is not likely to assist the reaction. Only the second term of the rate equations of problem 12.6 is significant here, with phosphine playing the role of the entering ligand.

12.13 Two factors need to be considered:

1. The best π acceptors (CO, PPh_3) slow the dissociation of CO in the *cis* position; π donors (halides) speed the dissociation.

2. There are fewer electrons available for π back-bonding in the halide complexes, resulting in weaker Cr–CO bonds, in spite of the π donation by the halides.

In addition, the complexes with very slow rates are d^6 species; the faster reacting complexes are d^5 species. All are low spin because of the large number of carbonyl ligands.

12.14 *Trans*-$Pt(NH_3)_2Cl_2$ reacts with thiourea (tu) to form *trans*-$[Pt(NH_3)_2(tu)_2]^{2+}$; the first Cl^- is easily removed (*trans* effect and inherent lability) and the strong *trans* effect of tu then helps replace the second Cl^-.

Both chlorides of *cis*-$Pt(NH_3)_2Cl_2$ are easily replaced (inherent lability of Cl). The tu's are then *trans* to the NH_3's, which are replaced because of the strong *trans* effect of tu, resulting in $[Pt(tu)_4]^{2+}$.

12.15 **a.** $[Pt(CO)Cl_3]^- + NH_3 \rightarrow$ *trans*-$[Pt(CO)(NH_3)Cl_2]$ CO is the stronger *trans* director.

b. $[Pt(NH_3)Br_3]^- + NH_3 \rightarrow$ *cis*-$[Pt(NH_3)_2Br_2]$ Br is the stronger *trans* director.

c. $[Pt(C_2H_4)Cl_3]^- + NH_3 \rightarrow$ *trans*-$[Pt(C_2H_4)(NH_3)Cl_2]$ C_2H_4 is the stronger *trans* director.

12.16 **a.** Two sets of reactions, with examples from Figure 12.12 identified:

$[PtCl_4]^{2-} + 2NH_3 \rightarrow$ *cis*-$[PtCl_2(NH_3)_2] + 2\ Cl^-$ (b)

cis-$[PtCl_2(NH_3)_2] + 2\ py \rightarrow$ *cis*-$[Pt(py)_2(NH_3)_2]^{2+} + 2\ Cl^-$ (h)

cis-$[Pt(py)_2(NH_3)_2]^{2+} + 2\ NO_2^- \rightarrow$ *trans*-$[Pt(NO_2)_2(NH_3)(py)] + NH_3 + py$ (e)

trans-$[Pt(NO_2)_2(NH_3)(py)] + CH_3NH_2 \rightarrow$
$[Pt<(NO_2)(CH_3NH_2)><(NH_3)(py)>]^+ + NO_2^-$ (g)

and

$[PtCl_4]^{2-} + 2\ py \rightarrow$ *cis*-$[PtCl_2(py)_2] + 2\ Cl^-$ (b)

cis-$[PtCl_2(py)_2] + 2\ CH_3NH_2 \rightarrow$ *cis*-$[Pt(CH_3NH_2)_2(py)_2]^{2+} + 2\ Cl^-$ (h)

cis-$[Pt(CH_3NH_2)_2(py)_2]^{2+} + 2\ NO_2^- \rightarrow$
trans-$[Pt(NO_2)_2(CH_3NH_2)(py)] + py + CH_3NH_2$ (c)

trans-$[Pt(NO_2)_2(CH_3NH_2)(py)] + NH_3 \rightarrow$
$[Pt<(NO_2)(NH_3)><(CH_3NH_2)(py)>]^+ + NO_2^-$ (a)

b. Reaction with Cl_2 puts Cl both above and below the plane, with the products

$$[Pt<(NO_2)(NH_3)><(CH_3NH_2)(py)><(Cl)(Cl)>]^+$$

and

$$[Pt<(NO_2)(CH_3NH_2)><(NH_3)(py)><(Cl)(Cl)>]^+$$

Reaction with one mole of Br^- replaces one Cl^-, with the products

$$[Pt<(NO_2)(NH_3)><(CH_3NH_2)(py)><(Br)(Cl)>]^+$$

and

$$[Pt<(NO_2)(CH_3NH_2)><(NH_3)(py)><(Br)(Cl)>]^+$$

12.17 **a.** The large negative entropy of activation implies that the activated complex is much more ordered than the reactants. This suggests an associative pathway.

b. The iodo ligand leads to the fastest rate, involving bond breaking *trans* to the halogen, and therefore has the strongest *trans* effect.

12.18 V^{2+} is d^3, and likely to be inert. V^{3+} is d^2, and labile. In low $[H^+]$ (or slightly basic) solutions, V^{3+} becomes VOH^{2+}, which may substitute into V^{2+} more readily. The V–OH–V combination must be significant in the rate, with a and b determining the relative rates of the two paths. Both are probably inner sphere, but we have no direct evidence, just similarity to Cr^{2+}–Cr^{3+} reactions. The rate equation is

$$\text{Rate} = (k_1 + k_2/[H^+])[V^{2+}][V^{3+}], \text{ so } a = k_1\,[V^{2+}][V^{3+}] \text{ and } b = k_2[V^{2+}][V^{3+}]$$

12.19 $[Cr(H_2O)_6]^{2+}$ is labile, but $[Co(NH_3)_6]^{3+}$ is inert, and the NH_3's cannot bridge readily (no way for them to bond with the Cr). Therefore, the reaction is likely to be an outer-sphere reaction.

12.20 This reaction is inner sphere, so the X^- acts as a bridging ligand. The order of rates is $Br^- > Cl^- > N_3^- > F^- > NCS^-$. The rate may be based partly on the substitution rates on Cr^{2+}, but that is likely to be a small effect. Most of the difference is the difference in transmission of the electron through the ligand and transfer of the ligand to the other Cr. Br^- works best, since it is soft, but not too soft. F^- is too hard, and holds the electrons too well; NCS^- has the soft end available for bonding to Cr^{2+}, which is hard. Based on structure, azide should be even faster, since it contains hard N and multiple bonds that usually transfer electrons readily. Perhaps the multiple bonds soften it, but this is only a lower limit on the rate. The data in this problem are from D.L. Ball and E. L. King, *J. Am. Chem. Soc.*, **1958,** *80*, 1091.

12.21 **a.** NSe has the stronger *trans* effect. The longer Os—N_1 distance is consistent with weakening of this bond by the ligand *trans* to it, the NSe ligand.

b. The short bond distance and large Os—N—Se angle (164.7°) suggest that the ligand has NSe^+ character; the cation NSe^+ would have a bond order of 3 and a very short bond distance. It is also worth noting that the N–Se stretching vibration in this complex is much higher (by 211 cm^{-1}) than in gas phase NSe.

12.22 **a.** The more rapid reactivity of the Mn complex is consistent with the general observation that first row transition metal complexes are generally more substitutionally labile than second and third row complexes.

b. The negative volume of activation implies that the activated complex is more highly ordered than the reactants, consistent with an *A* (or I_a) mechanism.

c. Occurrence of two infrared bands is more consistent with a *fac* isomer; a *mer* isomer is predicted by symmetry to have three IR-active bands:

mer isomer (C_{2v}):

C_{2v}	E	C_2	$\sigma(xz)$	$\sigma(yz)$	
Γ	3	1	3	1	
A_1	1	1	1	1	z
B_1	1	–1	1	–1	x

The representation Γ reduces to $2A_1 + B_1$, all IR-active. Three IR-active carbonyl bands are expected.

fac isomer (C_{3v}):

C_{3v}	E	$2C_3$	$3\sigma_v$	
Γ	3	0	1	
A_1	1	1	1	z
E	2	–1	0	(x, y)

The representation Γ reduces to $A_1 + E$, both IR-active. Two IR-active carbonyl bands are expected.

Chapter 13: Organometallic Chemistry

13.1 A: = method A, B: = method B

a.	$Fe(CO)_5$	A: 8 + 5×2 = 18	B: 8 + 5×2 = 18
b.	$[Rh(bipy)_2Cl]^+$	A: 7 + 2×4 + 2 = 17	B: 9 + 2×4 + 1 – 1 = 17
c.	$(\eta^5\text{-}Cp^*)Re(=O)_3$	A: 6 + 0 + 3×4 = 18	B: 5 + 7 + 3×2 = 18
d.	$Re(PPh_3)_2Cl_2N$	A: 2 + 2×2 + 2×2 + 6 = 16	B: 7 + 2×2 + 2×1 + 3 = 16
e.	$Os(CO)(\equiv CPh)(PPh_3)_2Cl$	A: 7 + 2 + 3 + 2×2 + 2 = 18	B: 8 + 2 + 3 + 2×2 + 1 = 18

13.2 All of these compounds have 16-electron valence configurations.

a.	$Ir(CO)Cl(PPh_3)_2$	A: 8 + 2 + 2 + 2×2 = 16	B: 9 + 2 + 1 + 2×2 = 16
b.	$RhCl(PPh_3)_3$	A: 8 + 2 + 3×2 = 16	B: 9 + 1 + 3×2 = 16
c.	$[Ni(CN)_4]^{2-}$	A: 8 + 4×2 = 16	B: 10 + 4×1 + 2 = 16
d.	*cis*-$PtCl_2(NH_3)_2$	A: 8 + 2×2 + 2×2 = 16	B: 10 + 2×2 + 2×1 = 16

13.3

a. $[M(CO)_7]^+$ — A: 18 – 7×2 = 4 = M^+, **V** — B: 18 – 7×2 = 4 = M^+, **V**

b. $H_3CM(CO)_5$ — A: 18 – 2 – 5×2 = 6 = M^+, **Mn** — B: 18 – 1 – 5×2 = 7 = M, **Mn**

c. $M(CO)_2(CS)(PPh_3)Br$ — A: 18 – 2×2 – 2 – 2 – 2 = 8 = M^+, **Co**
B: 18 – 2×2 – 2 – 2 – 1 = 9 = M, **Co**

d. $[(\eta^3\text{-}C_3H_3)(\eta^5\text{-}C_5H_5)M(CO)]^-$ — A: 18 – 4 – 6 – 2 = 6 = M^+, **Mn**
B: 18 – 3 – 5 – 2 = 8 = M^-, **Mn**

e. $(OC)_5M{=}C(OCH_3)C_6H_5$ — A and B: 18 – 5×2 – 2 = 6 = M, **Cr**

f. $[(\eta^4\text{-}C_4H_4)(\eta^5\text{-}C_5H_5)M]^+$ — A: 18 – 4 – 6 = 8 = M^{2+}, **Ni** — B: 18 – 4 – 5 = 9 = M^+, **Ni**

g. $(\eta^3\text{-}C_3H_5)(\eta^5\text{-}C_5H_5)M(CH_3)(NO)$

linear NO: A: 18 – 2 – 6 – 2 – 2 = 6 = M, **Cr**
B: 18 – 3 – 5 – 1 – 3 = 6 = M, **Cr**

bent NO: A: 18 – 2 – 6 – 2 – 2 = 6 = M^{2+}, **Fe**
B: 18 – 3 – 5 – 1 – 1 = 8 = M, **Fe**

h. $[M(CO)_4I(diphos)]^-$ — A: 18 – 4×2 – 2 – 4 = 4 = M, **Ti**
B: 18 – 4×2 –1 – 4 = 5 = M^-, **Ti**

13.4 Calculating for each metal atom:

a. $[Fe(CO)_2(\eta^5\text{-}C_5H_5)]_2$ A: $7 + 2\times2 + 6 = 17$, single Fe–Fe
B: $8 + 2\times2 + 5 = 17$, single Fe–Fe

b. $[Mo(CO)_2(\eta^5\text{-}Cp)]_2^{2-}$ A: $5 + 2\times2 + 6 + 1 = 16$, double Mo=Mo
B: $6 + 2\times2 + 5 + 1 = 16$, double Mo=Mo

13.5 **a.** $[M(CO)_3(NO)]^-$ linear M–N–O: A: $18 - 3\times2 - 2 - 2 = 8 =$ M, **Ru**
B: $18 - 3\times2 - 3 = 9 = M^-$, **Ru**
bent M–N–O: A: $18 - 3\times2 - 2 = 10 =$ M, **Pd**
B: $18 - 3\times2 - 1 = 11 = M^-$, **Pd**

b. $[M(PF_3)_2(NO)_2]^+$ linear NO: A: $18 - 2\times2 - 2\times2 = 10 = M^-$, **Rh**
B: $18 - 2\times2 - 2\times3 = 8 = M^+$, **Rh**

c. $[M(CO)_4(\mu_2\text{–}H)]_3$ As a triangular structure with three M–M bonds:
A: $18 - 4\times2 - 2 - 2 = 6 = M^+$, **Tc**
B: $18 - 4\times2 - 1 - 2 = 7 =$ M, **Tc**

d. $M(CO)(PMe_3)_2Cl$ A: $16 - 2 - 2\times2 - 2 = 8 = M^+$, **Rh**
B: $16 - 2 - 2\times2 - 1 = 9 =$ M, **Rh**

13.6 Method B works better for calculating overall charge.

a. $[Co(CO)_3]^z$ $9 + 3\times2 = 15, z = 3-$

b. $[Ni(CO)_3(NO)]^z$ $10 + 3\times2 + 3 = 19, z = 1+$

c. $[Ru(CO)_4(GeMe_3)]^z$ $8 + 4\times2 + 1 = 17, z = 1-$

d. $[(\eta^3\text{-}C_3H_5)V(CNCH_3)_5]^z$ $3 + 5 + 5\times2 = 18, z = 0$

e. $[(\eta^5\text{-}C_5H_5)Fe(CO)_3]^z$ $5 + 8 + 3\times2 = 19, z = 1+$

f. $[(\eta^5\text{-}C_5H_5)_3Ni_3(\mu_3\text{–}CO)_2]^z$ $3\times5 + 3\times10 + 2\times2 = 49, z = 1+$, assuming three Ni–Ni bonds; calculating for each Ni: $5 + 10 + 2(2/3) + 2 = 18\ 1/3$; charge per Ni = 1/3+, overall charge = 1+
(Each triply bridging CO can be considered to donate 2 electrons overall, 2/3 electron to each metal.)

13.7 **a.** $[(\eta^5\text{-}C_5H_5)W(CO)_x]_2$, assuming a single W–W:
A: $6 + 5 + 1 + x\times2 = 18, x = 3$
B: $5 + 6 + 1 + x\times2 = 18, x = 3$

b. $ReBr(CO)_x(CO_2C_2H_4)$ A: $6 + 2 + x\times2 + 2 = 18, x = 4$
B: $7 + 1 + x\times2 + 2 = 18, x = 4$

c. $[(CO)_3Ni\text{–}Co(CO)_3]^z$ A and B: $3\times2 + 10 + 2 + 9 + 3\times2 - z = 36, z = 3-$

d. $[Ni(NO)_3(SiMe_3)]^z$ B: $10 + 3\times3 + 1 - z = 18, z = 2+$

e. $[(\eta^5\text{-}C_5H_5)Mn(CO)_x]_2$ B: $5 + 7 + 2 + x \times 2 = 18, x = 2$

13.8 **a.**

Upper ring: η^5-Cp	5 electrons
Co	9 electrons
	14 electrons; 4 more needed, so lower ring must be η^4

b.

M:	η^1-C_7 ring	1 electron	**M′**: η^7-C_7 ring	7 electrons
	2 PEt_3	4 electrons	η^5-C_5 ring	5 electrons
	GeR_3	1 electron		12 electrons
		6 electrons		4 more needed to reach 16: Ti
	10 more needed to reach 16: Pt			

13.9 Figure 10.20 (page 389) gives an MO diagram for $Ni(CO)_4$ and cites additional references. The HOMOs (t_2) are strongly bonding, and there is a large energy gap between the HOMOs and the LUMOs in this 18-electron molecule.

13.10 The energy of stretching vibrations depends on the square root of the force constant divided by the reduced mass (Section 13.4.1).

The reduced masses are 14.73 for ^{16}O and 1.41 for ^{18}O, so $(14.73/16.41)^{1/2} = 0.947$, and the ^{18}O complex has vibrational energy of $0.947 \times 975\ cm^{-1} = 924\ cm^{-1}$. The value given in the reference is $926\ cm^{-1}$.

13.11 Sulfur is less electronegative than oxygen. Therefore, the tungsten in $W(S)Cl_2(CO)(PMePh_2)_2$ has greater electron density and a greater tendency to back-donate to Co; a lower energy ν(CO) is expected (actual value: $1986\ cm^{-1}$).

13.12 Adding electrons to a carbonyl complex puts more electrons into the back-bonding orbitals. As a result, the V–C bonds in $[V(CO)_6]^-$ are strengthened and the distance shortened (but the C–O bonds are weakened by having more electrons in the antibonding orbitals).

13.13

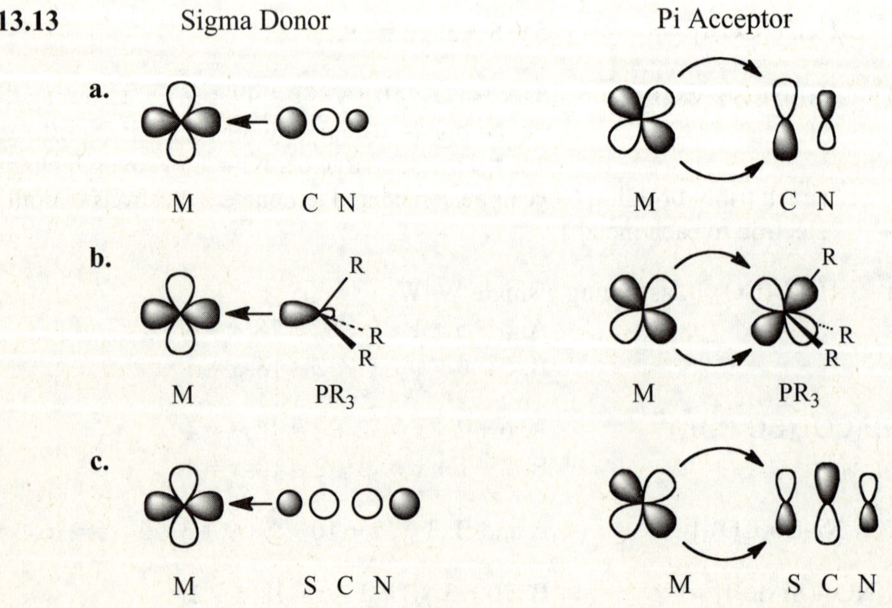

13.14 a. If NO is counted as a linear donor, each of these has 18 electrons. The increasing nuclear charge draws the NO electrons in toward the metal, resulting in less backbonding to the ligands and an increase in N–O bond order. The Cr species has an energy below that of most linear species, so it may have bent NO coordination, making it a 16-electron ion. The Fe complex has an energy above the normal, so its π back-bonding must be unusually weak.

b. The low energy band indicates bent NO coordination; the higher energy band is from the linear ligand. This complex has one of each, with angles of 138° and 178° (Greenwood and Earnshaw, *Chemistry of the Elements*, 2nd ed., pages 450–452).

13.15 a. The CO_2 molcular orbitals are shown in Figure 5.26 on page 157.

b. The π orbitals of 1,3,5-hexatriene are shown in Figure 13.21 on page 510.

c. *cyclo*-C_4H_4:

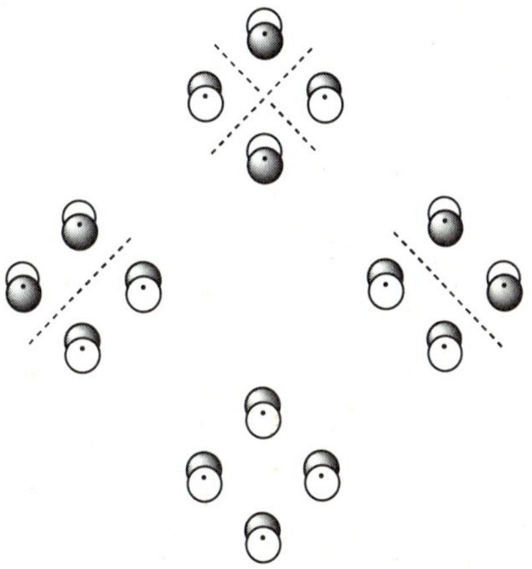

d. *cyclo*-C_7H_7:

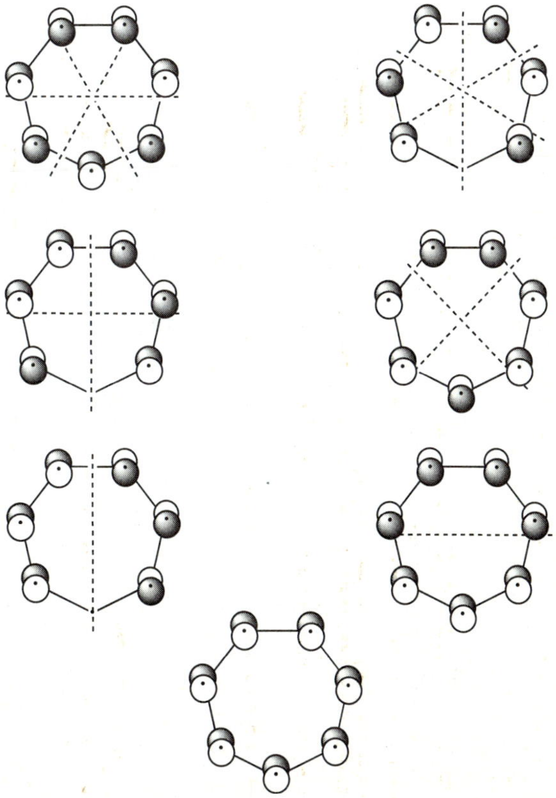

13.16 **a.** Using the group theoretical method described in Chapter 4: The four CO ligands are in a square planar arrangement, but the Mo atom is above them. The reducible representation Γ, shown below, can be derived for the C–O stretching vibrations. Γ reduces to $A_1 + B_1 + E$, with A_1 and E infrared active, so there are two bands visible (the E bands are degenerate).

C_{4v}	E	$2C_4$	C_2	$2\sigma_v$	$2\sigma_d$	
Γ	4	0	0	2	0	
A_1	1	1	1	1	1	z
B_1	1	–1	1	1	–1	
E	2	0	–2	0	0	(x, y)

b.

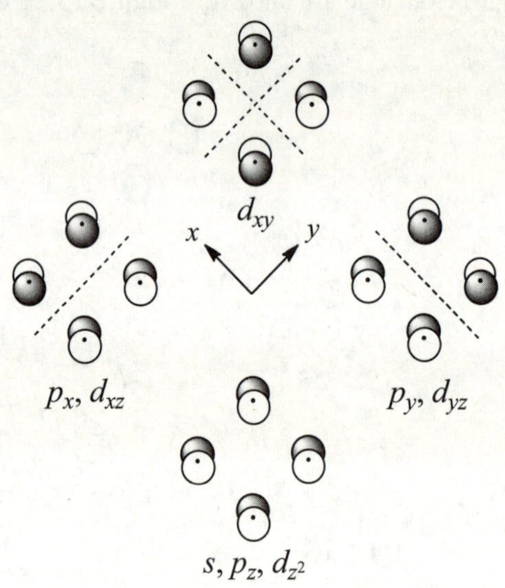

13.17 **a.** From bottom to top:

0-node, left, is A_1', 0 node, right is A_2''
1-node, far left and third from left are E_1'
1-node, second from left and far right are E_1''
2-node, far left and third far left are E_2'
2-node, second from left and far right are E_2''

b.

s, d_{z^2}	$(d_{xy}, d_{x^2-y^2})$	(d_{xy}, d_{yz})	p_z	(p_x, p_y)
A_1'	E_2'	E_1''	A_2''	E_1'

c. The matching orbitals are shown in part b; there is no match in the Fe orbitals for the E_2'' ligand orbitals.

13.18 **a.** The π orbitals of benzene are shown in Figure 13.22, page 511.

b. Group orbitals

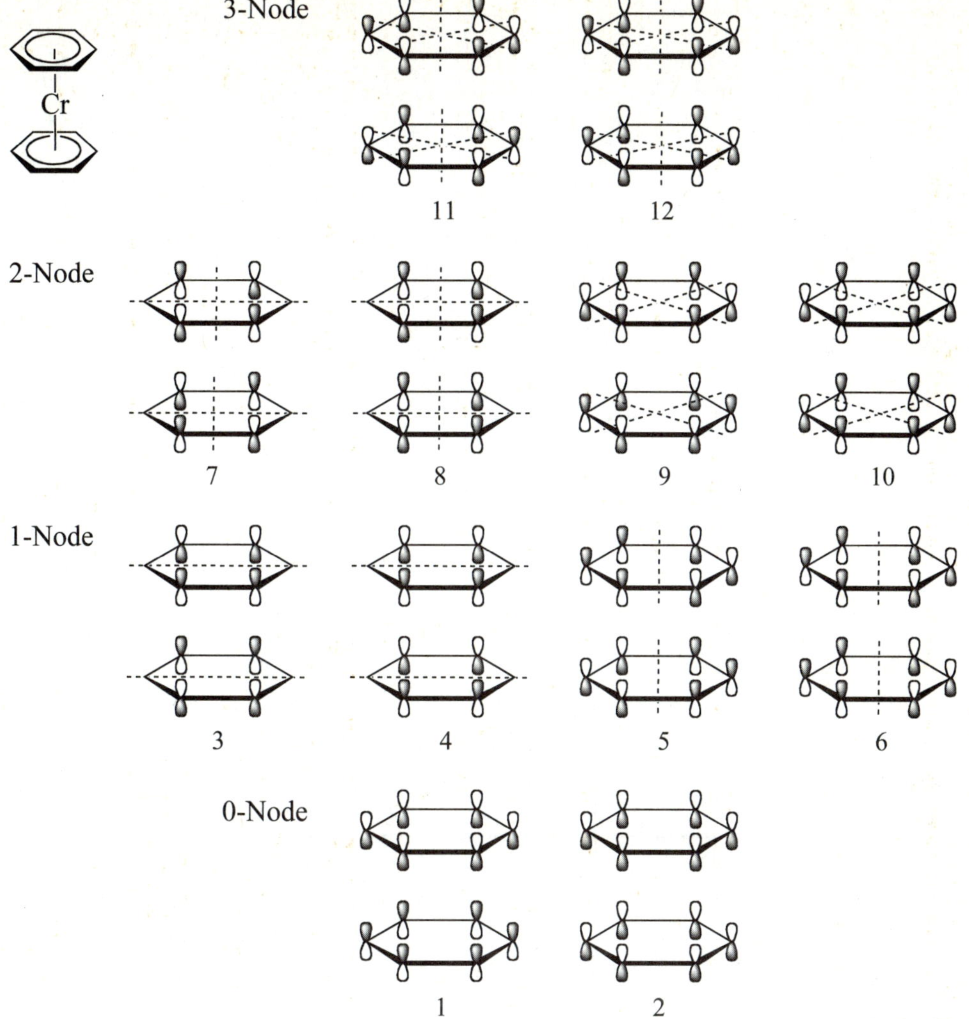

c. Matching Cr orbitals:

1	s, d_{z^2}	7	d_{xy}
2	p_z	8	none
3	p_y	9	$d_{x^2-y^2}$
4	d_{yz}	10	none
5	p_x	11	none
6	d_{xz}	12	none

d. Energy level diagram

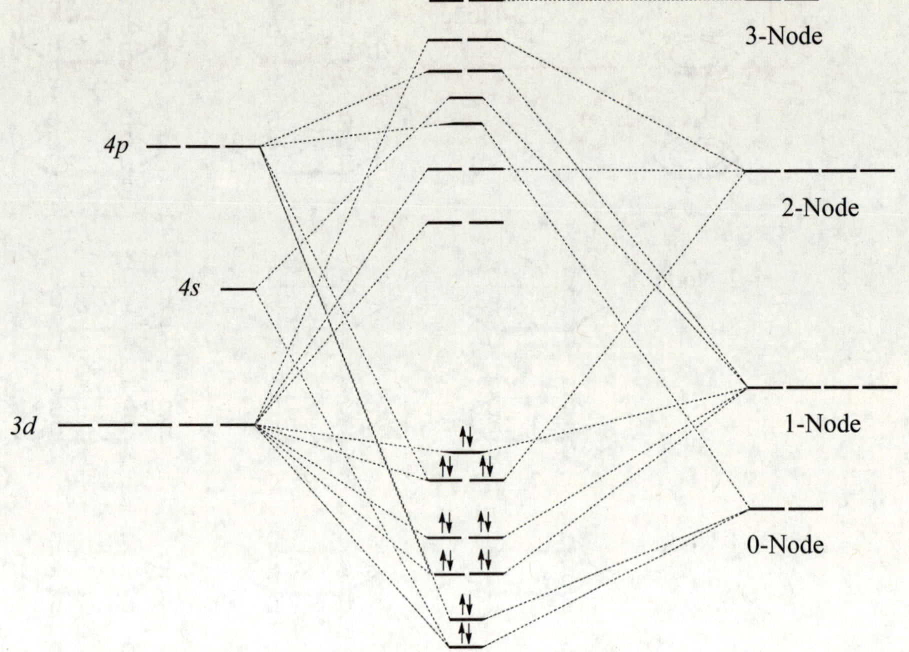

13.19

C_{3v}	E	$2C_3$	$3\sigma_v$	
Γ	3	0	1	
A_1	1	1	1	z
A_2	1	1	–1	
E	2	–1	0	(x, y)

$\Gamma = A_1 + E$, both IR active. Three vibrations, but two are degenerate, so two C–O stretching bands are expected in the IR spectrum.

13.20 $Ni(CO)_4$

T_d	E	$8C_3$	$3C_2$	$6S_4$	$6\sigma_d$	
Γ	4	1	0	0	1	
A_1	1	1	1	1	1	
T_2	3	0	–1	–1	1	(x, y, z)

$\Gamma = A_1 + T_2$ A_1 is IR inactive, T_2 is IR active: 1 band.

$Cr(CO)_6$

O_h	E	$8C_3$	$6C_2$	$6C_4$	$3C_2$	i	$6S_4$	$8S_6$	$3\sigma_h$	$6\sigma_d$	
Γ	6	0	0	2	2	0	0	0	4	2	
A_{1g}	1	1	1	1	1	1	1	1	1	1	
E_g	2	−1	0	0	2	2	0	−1	2	0	
T_{1u}	3	0	−1	1	−1	−3	−1	0	1	1	(x, y, z)

$\Gamma = A_{1g} + E_g + T_{1u}$ Only T_{1u} is IR active: 1 band

13.21

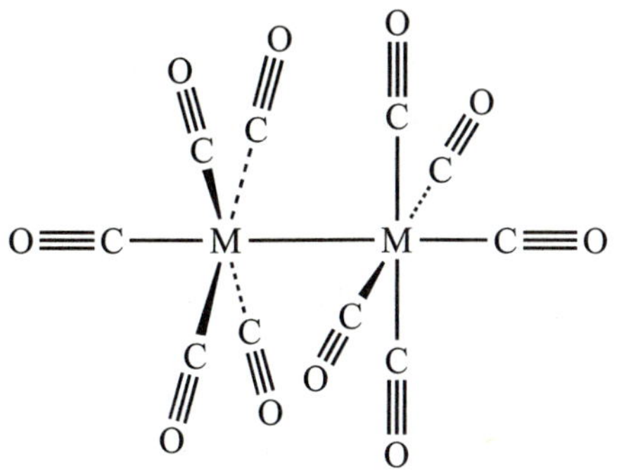

D_{4d}	E	$2S_8$	$2C_4$	$2S_8^3$	C_2	$4C_2'$	$4\sigma_d$	
Γ	10	0	2	0	2	0	4	
A_1	1	1	1	1	1	1	1	
B_2	1	−1	1	−1	1	−1	1	z
E_1	2	$\sqrt{2}$	0	$-\sqrt{2}$	−2	0	0	(x, y)
E_2	2	0	−2	0	2	0	0	
E_3	2	$-\sqrt{2}$	0	$\sqrt{2}$	−2	0	0	(R_x, R_y)

$\Gamma = 2A_1 + 2B_2 + E_1 + E_2 + E_3$ $2B_2$ and E_1 are IR active: 3 bands

13.22 a.

$W(CO)_5(NCC_3H_7)$ 2077, 1975, 1938 cm^{-1}
$W(CO)_4(NCC_3H_7)_2$, *cis* 2107, 1898, 1842 cm^{-1}
$W(CO)_3(NCC_3H_7)_3$, probably *fac* 1910, 1792 cm^{-1}

Pentacarbonyls have three active IR stretches, tetracarbonyls have 1 (other ligands *trans*) or 4 (other ligands *cis*), and tricarbonyls have 2 (*fac*) or 3 (*mer*) (see Table 13.7). It appears that two of the tetracarbonyl bands have similar enough energies for their bands to overlap, so only 3 appear in the spectrum (or one band may be so weak in intensity that it does not appear), but that the nitriles must be *cis*. The evidence for the *fac* isomer of the tricarbonyl is not conclusive based on the IR spectrum alone, since again two bands of the *mer* isomer might overlap.

b. The stretching energy of the CO *trans* to the nitrile ligand is lower than that for CO *cis* due to reduced competition for the π back-bonding electrons (CO is a better π-acceptor). In the second complex, two are *cis* and two are *trans*. In the third complex, all three are *trans*, so the band energy is lower. In general, the trend is to lower energies as CO is replaced by butyronitrile; butyronitrile is not as effective a π acceptor as CO, so the π-acceptor nature of CO is enhanced as more nitrile ligands are added.

c. As more CO ligands are replaced by nitriles, which are stronger at donating electrons to Mo, the remaining CO ligands become stronger π acceptors, and the Mo–C bonds become stronger. In $Mo(CO)_3(NCC_3H_7)_3$ the Mo–C bond is so strong that the complex cannot react further with butyronitrile.

13.23 a. The bands at lower energy are for the CO ligands. The C–O stretches, which involve a greater change in dipole moment, are more intense than the C–N stretches. Because CO has a slightly greater reduced mass than CN, C–O stretches should occur at lower energies than C–N stretches; the energy of vibrational levels is proportional to $\sqrt{\text{reduced mass}}$. Also, the greater π-acceptor ability of the CO ligand is a contributing factor to the lower energy stretches for this ligand.

b. The *trans*-$[Fe(CO)_2(CN)_4]^{2-}$ complex should have a single IR-active C–O stretch (only the antisymmetric stretch is IR active) and a single IR-active C–N stretch (see Table 13.7). The *cis* complex should have two IR-active C–O stretches (both symmetric and antisymmetric) and four IR-active C–N stretches. In addition to its single carbonyl stretch, $[Fe(CO)(CN)_5]^{3-}$ would be expected to have three IR-active C–N stretches. The correct identifications are therefore:

	Predicted on basis of symmetry	
Complex	C–O stretches	C–N stretches
A: *cis*-$[Fe(CO)_2(CN)_4]^{2-}$	2	4
B: *trans*-$[Fe(CO)_2(CN)_4]^{2-}$	1	1
C: $[Fe(CO)(CN)_5]^{3-}$	1	3

In complexes **A** and **C** some of the C–N stretches are too weak to be seen or the bands overlap; otherwise, the expected numbers of bands and the observed spectra match.

13.24 a. The CO ligand can be in either the axial or the equatorial position on the trigonal bypyramidal complex. The electronic environments are not identical, so the two isomers should have different carbonyl stretching bands.

b. The CO stretch absorbs at a higher energy in $Fe(CO)(PF_3)_4$, implying that the C–O bond is stronger in that compound. If the CO bond is stronger, the Fe–C bond is correspondingly weaker, indicating that the PF_3 ligands are better π acceptors and therefore higher in the spectrochemical series than CO.

13.25

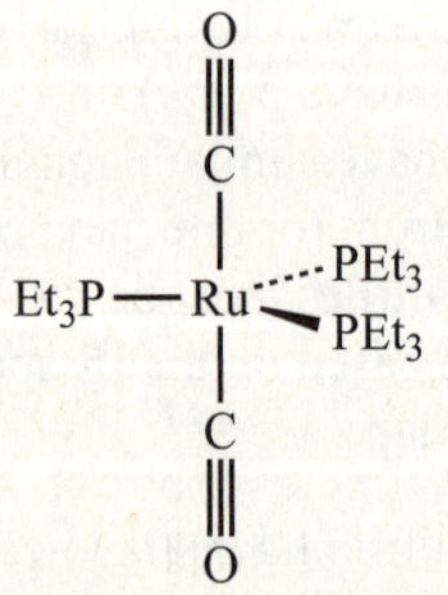

One C–O stretch (antisymmetric) (minor isomer)

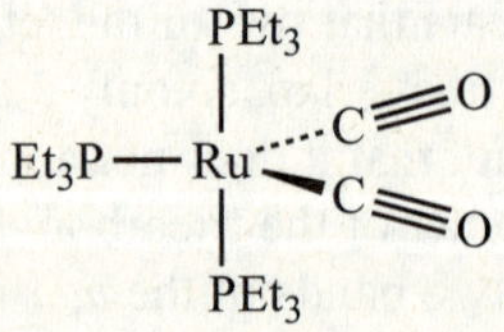

Two C–O stretches (symmetric and antisymmetric) (major isomer)

13.26 From Table 13.7 (page 537), $[Co(CO)_3(PPh_3)_2]^+$ must be trigonal bipyramidal with all three CO ligands in the equatorial plane, giving rise to a single C–O absorption band.

13.27 By extrapolation of the positions of the C–O bands for $[Mn(CO)_6]^+$ (2100 cm^{-1}) and $[Fe(CO)_6]^{2+}$ (2204 cm^{-1}), one might predict a comparable band for $[Ir(CO)_6]^{3+}$ near 2300 cm^{-1}. Actual value: 2254 cm^{-1}.

13.28 **a.** In all three cations the strongly positive metal pulls electrons strongly from the CO ligands, significantly reducing back-bonding. This effect is greatest for $[Hg(CO)_2]^{2+}$, with a charge of 1+ per CO, and weakest for $[Os(CO)_6]^{2+}$, with a charge of 1/3+ per CO. Therefore, $[Os(CO)_6]^{2+}$ should have the strongest back-bonding, weakest C–O bond, and lowest energy carbon-oxygen stretching vibration. Actual positions of IR bands:

	ν(CO), cm^{-1}
$[Hg(CO)_2]^{2+}$	2278
$[Pt(CO)_4]^{2+}$	2244
$[Os(CO)_6]^{2+}$	2190

b. A reducible representation for this ion based on C–O stretching vibrations would have the following characters:

D_{2d}	E	$2S_4$	C_2	$2C_2'$	$2\sigma_d$	
Γ	6	0	2	0	4	
2 A_1	2	2	2	2	2	
2 B_2	2	–2	2	–2	2	z
E	2	0	–2	0	0	(x, y)

There should be three carbon–oxygen stretching bands, two of B_2 symmetry and one of E symmetry (a degenerate pair). With a charge 2+, this complex should exhibit infrared bands near those of $[Fe(CO)_6]^{2+}$ (2204 cm^{-1}). Actual bands are observed at 2173 cm^{-1} (E) and at 2187 and 2218 cm^{-1} (B_2).

13.29 **a.** The representation based on the set of six C–O vibrations in O_h symmetry is:

O_h	E	$8C_3$	$6C_2$	$6C_4$	$3C_2(=C_4^2)$	i	$6S_4$	$8S_6$	$3\sigma_h$	$6\sigma_d$
Γ	6	0	0	2	2	0	0	0	4	2

This representation reduces to $A_{1g} + E_g + T_{1u}$ (see Exercise 10.4, page 371). Of these, A_{1g} and E_g match squared functions and are therefore Raman active. (A_{1g} matches the band at 2015 cm^{-1} and E_g matches the band at 2119 cm^{-1}.)

b. **J**, **K**, and **L** are, respectively, the products of substitution of CO by py: $Mo(CO)_5py$, *cis*-$Mo(CO)_4py_2$, and *fac*-$Mo(CO)_3py_3$. As the number of pyridine ligands increases, the strong σ donation by this ligand increases the back-bonding by the CO ligands, resulting in successive lowering in energy of the C–O vibrations. Symmetry analysis of each of these complexes shows more expected Raman-active bands than reported in the article. Additional weak bands can be seen in the Raman spectra (shown in the reference).

13.30 $(\eta^5\text{-}C_5H_5)Cr(CO)_2(NS)$ 1962, 2033 cm^{-1}

$(\eta^5\text{-}C_5H_5)Cr(CO)_2(NO)$ 1955, 2028 cm^{-1}

NS is a stronger π acceptor, so the CO back-bonding is reduced in $(\eta^5\text{-}C_5H_5)Cr(CO)_2(NS)$ and the C≡O bond is strengthened and has a higher stretching energy.

13.31 The energy for a stretching vibration is proportional to $\sqrt{\frac{k}{\mu}}$, where k is the force constant and μ is the reduced mass (Section 13.4.1). In this case, the reduced masses for the N–S stretches are: ^{14}N–S: $\mu = \frac{14.00 \times 32.06}{14.00 + 32.06} = 9.745$ ^{15}N–S: $\mu = \frac{15.00 \times 32.06}{15.00 + 32.06} = 10.21$

Because the energy is inversely proportional to the square root of the reduced mass, we can write:

$\frac{E(^{15}N-S)}{E(^{14}N-S)} = \sqrt{\frac{\mu(^{14}NS)}{\mu(^{15}NS)}} = \sqrt{\frac{9.745}{10.21}} = 0.9770$ The expected position of the N–S stretch in the ^{15}NS complex is therefore 0.9770 × 1284 cm^{-1} = 1254 cm^{-1}. The reported value is 1248 cm^{-1}.

13.32 Because the reduced mass of ^{13}CO is greater than the reduced mass of ^{12}CO, the separation between vibrational energy levels should be less for ^{13}CO, and the ^{13}CO complex should therefore show an infrared band at lower energy than 2199 cm^{-1}. Actual value: 2149 cm^{-1}.

13.33 **a.** $Mo(CO)_6 + Ph_2PCH_2PPh_2 \longrightarrow H_2C(PPh_2)_2Mo(CO)_4 + 2\,CO$

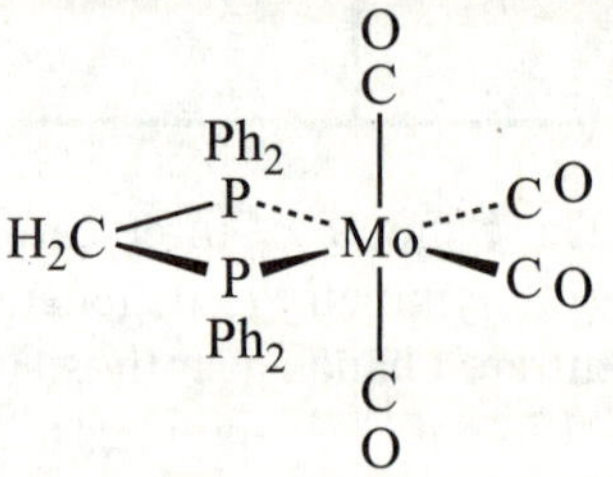

b. $(\eta^5\text{-}C_5H_5)(\eta^1\text{-}C_3H_5)Fe(CO)_2 \xrightarrow{h\nu} (\eta^5\text{-}C_5H_5)(\eta^3\text{-}C_3H_5)Fe(CO) + CO$

The allyl ligand can bond in either η^1 or η^3 fashion. Loss of CO converts the reactant from 18 electrons to 16 electrons; rearrangement of allyl from η^1 to η^3 returns it to 18 electrons.

c. $(\eta^5\text{-}C_5Me_5)Rh(CO)_2 \longrightarrow [(\eta^5\text{-}C_5Me_5)Rh(CO)]_2 + 2\,CO$

A double Rh–Rh bond is needed; CO ligands may be bridging or terminal (the electron count is the same for both modes).

d. $V(CO)_6 + NO \longrightarrow V(CO)_5(NO) + CO$

The compound changes from 17 to 18 electrons.

e. $W(CO)_5{=}C(C_6H_5)(OC_2H_5) + BF_3 \longrightarrow [(CO)_5W{\equiv}CC_6H_5]^+ + F^- + F_2BOC_2H_5$

f. $[(\eta^5\text{-}C_5H_5)Fe(CO)_2]_2 + 2\ Al(C_2H_5)_3 \longrightarrow$

(See Figure 13.17)

13.34

	ν(CO), cm^{-1}
$P(t\text{-}C_4H_9)_3$	1923 (best σ donor, poorest π acceptor)
$P(p\text{-}C_6H_4Me)_3$	1965
$P(p\text{-}C_6H_4F)_3$	1984
$P(C_6F_5)_3$	2004 (best π acceptor)

13.35 **a.** $Fe(CO)_4(PF_3)$ (PF_3 is a strong π acceptor)

b. $[Re(CO)_6]^+$

c. $Mo(CO)_3(PCl_3)_3$ (PCl_3 is the best π acceptor among the phosphines)

13.36 In order from highest to lowest ν:

$Mo(CO)_4\,(F_2PCH_2CH_2PF_2)$	has strongest acceptor diphosphine
$Mo(CO)_4\,((C_2F_5)_2PCH_2CH_2P(C_2F_5)_2)$	
$Mo(CO)_4\,((C_6F_5)_2PCH_2CH_2P(C_6F_5)_2)$	
$Mo(CO)_4\,(Ph_2PCH_2CH_2PPh_2)$	
$Mo(CO)_4\,(Et_2PCH_2CH_2PEt_2)$	has strongest donor diphosphine

13.37 Coordinated N_2 has a lower stretching energy than free N_2. N_2 can act as a π acceptor, weakening the N–N bond and lowering the energy of the stretching vibration. The stretching vibration of free N_2 is not seen in the IR spectrum because there is no change in dipole moment on stretching.

13.38 At the higher temperature, the $-O-CH_3$ group should rotate rapidly enough to show only a single, average environment, so each CH_3 has a single peak. At low temperatures, this rotation can be restricted, and *cis* and *trans* isomers result (see Figure 13.45). The larger peaks represent the more prevalent isomer.

13.39 At high temperature the C_5H_5 rings undergo rapid 1,2 shifts to give a single average proton signal on the NMR time scale. At sufficiently low temperature, the different environments of the η^1 and η^5 rings can be seen. The relative intensities are a = 1, b = 2, c = 2, and d = 5.

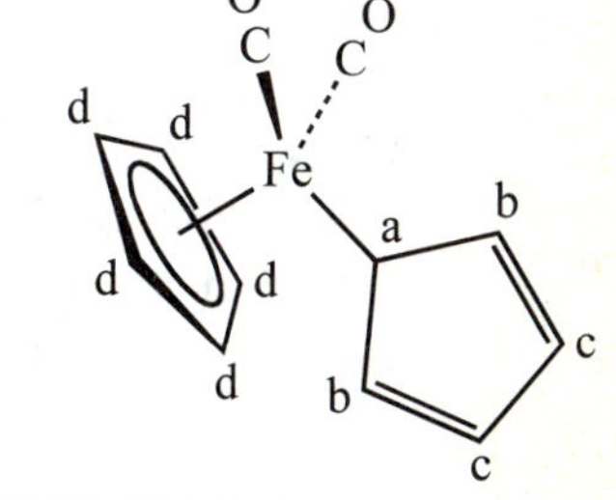

13.40 PF_3 is a stronger π acceptor than PCl_3. As a result, the chromium in $Cr(CO)_5(PCl_3)$ has a greater electron density, and CO acts as a stronger π acceptor in this complex.

a. $Cr(CO)_5(PF_3)$ has the stronger and shorter C–O bonds.

b. $Cr(CO)_5(PCl_3)$ has the higher energy Cr–C bands; since CO acts as a better π acceptor in this complex, the Cr–C bond is strengthened.

13.41 **a.** $[Fe(NO)(mnt)_2]^-$ has less electron density on Fe, less back-bonding to NO, and a stronger N–O bond with higher stretching frequency.

b. $(CO)_5Cr{:}N{\equiv}N{:}Cr(CO)_5$ N_2 acts as a π acceptor toward both metals, significantly weakening the N–N bond.

c. $Ta{=}CH_2$ has a shorter Ta–C bond because it is a double bond.

d. $Cr{\equiv}CCH_3$ has a triple bond, which is shorter than either of the Cr–C bonds to CO ligands.

e. $[Fe(CO)_4]^{2-}$ has more π back-bonding because Fe has the lowest nuclear charge of the metals in this isoelectronic series; this reduces the C≡O bonding and the energy of the C–O vibration.

13.42 One CO is replaced by 2-butyne. The NMR peaks are due to ethyl CH_3 ($\delta = 0.90$), ethyl CH_2 ($\delta = 1.63$), and butyne CH_3 ($\delta = 3.16$). The $\delta = 3.16$ peak splits at low temperatures because the two ends of the butyne are not identical; at higher temperatures, they become identical on NMR time scale, perhaps through rotation about the Mo–butyne bond. The single ^{31}P peak indicates identical PEt_3 groups, suggesting the isomer shown. The IR indicates that a CO ligand remains on the compound; the molecular weight of the compound shown is 574.2, well within the limits given.

13.43 The analysis fits $[(\eta^5\text{-}C_5H_5)Fe(\mu\text{-}CO)_3Fe(\eta^5\text{-}C_5H_5)]$. The single CO band is consistent with this structure. Using D_{3h} symmetry for the central part of the molecule, a representation based on the bridging carbonyls reduces to A_1' (IR-inactive) plus E' (IR-active); the observed absorption is for the E' vibration.

D_{3h}	E	$2C_3$	$3C_2$	σ_h	$2S_3$	$3\sigma_v$	
Γ	3	0	1	3	0	1	
A_1'	1	1	1	1	1	1	
E'	2	−1	0	2	−1	0	(x, y)

13.44 $(\eta^5\text{-}C_5H_5)(\eta^3\text{-}C_5H_7)Ni$ There are five η^5 protons, four on the two carbons of the second Cp ring that are not bonded to Ni, two on the first and third carbons bonded to Ni, and one on the center C bonded to Ni.

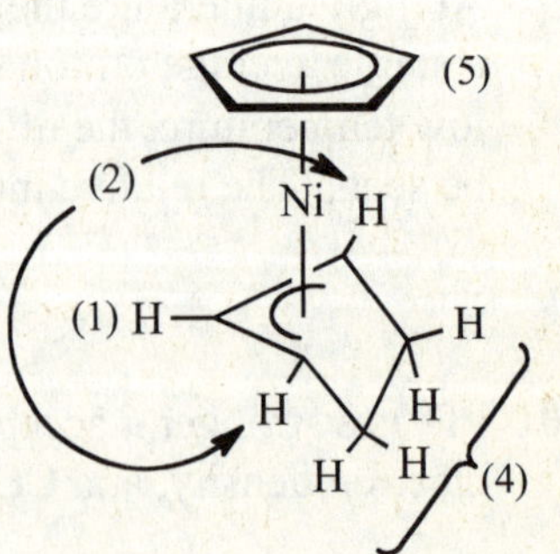

13.45 $I_s/I_a = \cot^2(\phi/2) = \cot^2 38° = 1.64$

13.46 a. In the η^6-C_6H_6 complex, CO is acting as a stronger π acceptor, as shown in its lower energy C–O stretching vibrations. Because the concentration of electrons on Cr is therefore greater in the η^6-C_6H_6 complex, η^6-C_6H_6 is donating more strongly to Cr than is the η^6-C_4BNH_6 ligand, so η^6-C_4BNH_6 is the stronger acceptor.

b. The longest C–C distance is the bond opposite the B–N bond, as in the resonance structure shown. The nitrogen attracts electrons from its neighboring carbon, which in turn attracts electrons from the next carbon, enhancing the bond strength between these two carbons and resulting in the shortest C–C bond, 1.374 Å. In the π orbitals (see Figure 13.22 for benzene), one of the occupied π orbitals is antibonding with respect to the C–C bond opposite the B–N bond, consistent with this C–C bond being the longest in the molecule.

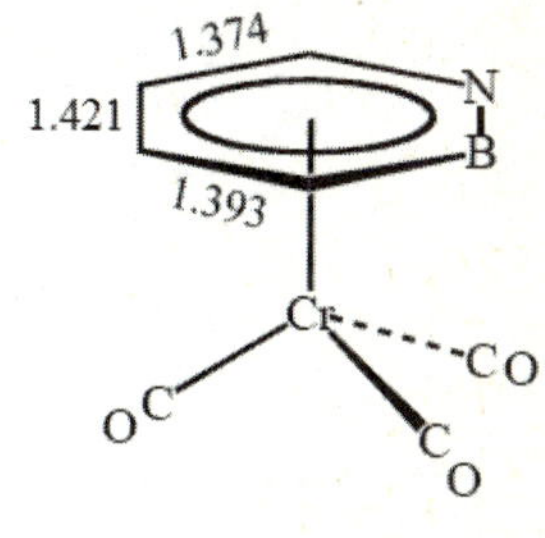

13.47 a.

Mass of C_{60}:	720
Ir:	193 (most abundant isotope)
CO:	28
C_9H_7:	115
	1056

b. Because the carbonyl stretch decreases by 44 cm^{-1}, there must be a significant decrease in the electron density at Ir.

c. The C_{60} is replaced by PPh_3, giving (η^5-C_9H_7)Ir(CO)(PPh_3).

13.48 Spectral data are interpreted in the reference.

13.49 If CO is liberated from (η^5-C_5H_5)Mn$(CO)_3$ it must be replaced by some other ligand if the 18-electron rule is to be maintained. In this case, the new ligand is tetrahydrofuran (THF), a cyclic ether that can act as a sigma donor, and compound **Q** is (η^5-C_5H_5)Mn$(CO)_2$(THF):

$$(\eta^5\text{-}C_5H_5)Mn(CO)_3 + THF \longrightarrow \underset{\mathbf{Q}}{(\eta^5\text{-}C_5H_5)Mn(CO)_2(THF)} + CO$$

The NC groups on carbon in $H_2C(NC)_2$ can act as donors to transition metals; each has a lone pair on carbon. In the formation of **R**, the weakly bound THF is replaced by a $H_2C(NC)_2$ ligand:

+ CpMn$(CO)_2$(THF) ⟶ **R**

13.50 There are 0.00300 mmol of dimers in the solution. Of the metals in these dimers, one third are Mo and two thirds are W.

The probability that a particular molecule has the formula $[CpMo(CO)_3]_2$ = 1/3 × 1/3 = 1/9. Therefore, 1/9 × total moles = 1/9 × 0.00300 mmol = 0.00033 mmol $[CpMo(CO)_3]_2$.

The probability that a molecule has the formula $[CpW(CO)_3]_2$ = 2/3 × 2/3 = 4/9, and 4/9 × 0.00300 mmol = 0.00133 mmol $[CpW(CO)_3]_2$.

The probability that a molecule has the formula $Cp(CO)_3Mo–W(CO)_3Cp$ = 2 × 1/3 × 2/3 = 4/9, and 4/9 × 0.00300 mmol = 0.00133 mmol $Cp(CO)_3Mo–W(CO)_3Cp$.

13.51 The IR bands at 1945 and 1811 cm^{-1} are similar to those of $[Ti(CO)_4(\eta^5\text{-}C_5H_5)]^-$, suggesting similarites in structure and charge between this Ti complex and **Z**. If the other IR bands are for B–H stretches and the ^{1}H NMR shows two signals of relative area 3:1, it is reasonable to suggest that the boron might be present in the BH_4^- ion, with three of the hydrogens in one environment, the fourth in another. The peak at 2495 cm^{-1} is distinctly different in energy from the peaks at 2132 and 2058 cm^{-1} and might correspond to a stretch involving the less abundant of the protons. All is consistent with a structure having 4 CO ligands and one BH_4^- ligand (which has replaced the cyclopentadienyl ligand), and a negative charge.

Symmetry analysis indicates that this complex should have two C–O stretching bands in the infrared (A_1 and E in local C_{4v} symmetry) and three B–H bands (two A_1 and one E in local C_{3v} symmetry). The higher energy B–H stretch is for the terminal H; the lower energy bands are for the bridging H atoms.

13.52 Elemental analysis: calculated: 36.4% C, 6.10% H by mass. ^{1}H NMR: peak at δ 2.02 corresponds to methyl groups on Cp ring (15 protons), peak at δ –11.00 corresponds to hydrides (which typically have negative chemical shifts) (5 protons); protons are in desired 3:1 ratio. IR: If the environment around the osmium is assigned C_{4v} symmetry, using the method of Chapter 4 gives the representation:

C_{4v}	E	$2C_4$	C_2	$2\sigma_v$	$2\sigma_d$
Γ	5	1	1	3	1

This reduces to $2A_1 + B_1 + E$. The A_1 and E representations are IR-active, for a total of three IR-active vibrational modes, matching the three bands in the spectrum.

13.53 Different software and different parameter settings will generate orbitals with slightly different shapes and energies than those shown in the text, although the results should be similar. The relative energies in part c in particular may differ from those in the text, because this calculation is sensitive to the methods used. In part d, the d_{z^2} conical nodal surface of the orbital is close to the p orbitals of the C_5H_5 rings (the d_{z^2} lobes point toward the centers of the rings), so the interaction between this orbital and the rings is weak.

13.54 Two orbitals that show ethylene as both a donor and acceptor are shown below.

Donor interaction (from π orbital of ethylene):

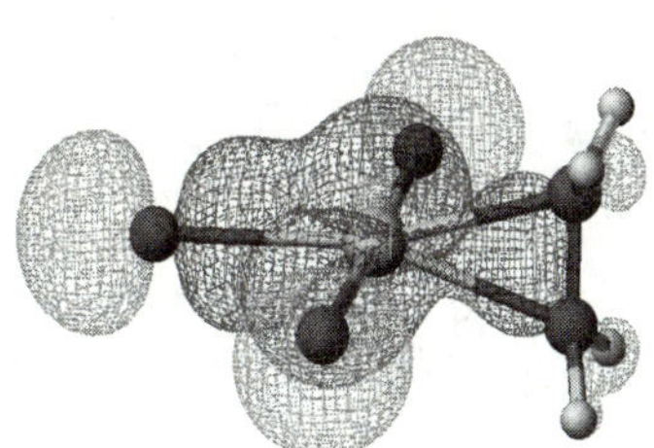

Acceptor interaction (involving π* orbital of ethylene):

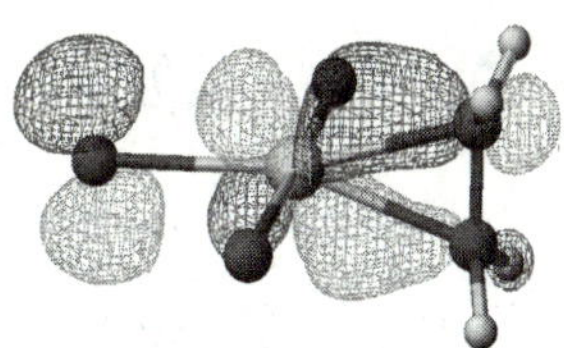

13.55 The shapes of the orbitals should be similar to those in Figure 13.8. The e_g* orbitals have different shapes because the d orbitals involved (d_{z^2} and $d_{x^2-y^2}$) have different shapes.

13.56 **a.** See orbitals in the middle of page 510.

b.

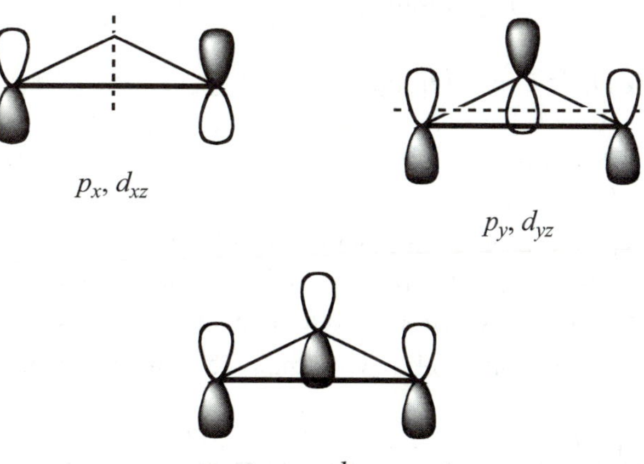

c. Depending on the parameters used, the upper and lower lobes of the p orbitals should merge in the lowest energy π orbital (to give clouds above and below the plane of the nuclei), and the lobes of two p orbitals in one of the upper π orbitals (derived from the p orbitals in front in the diagram at upper right above) may also merge.

d. The analysis can proceed similarly to the discussion of ferrocene on pages 514–518.

13.57 a.

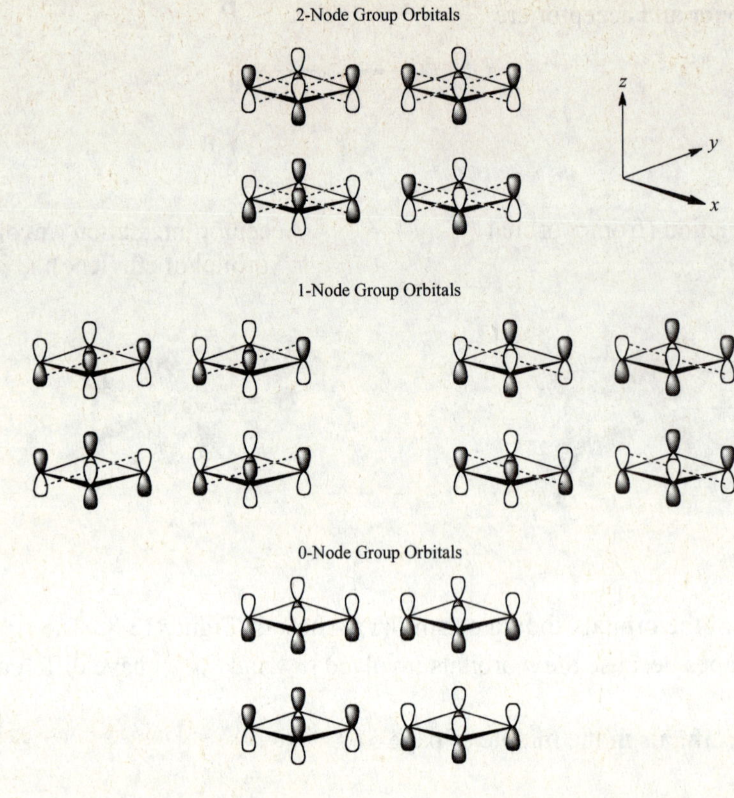

b.

2-Node Group Orbitals:		d_{xy}	none	
1-Node Group Orbitals:	p_x	d_{xz}	p_y	d_{yz}
0-Node Group Orbitals:		s, d_{z^2}	p_z	

c. The representation Γ, shown below, reduces to the irreducible representations listed. These match the group orbitals and their matching metal orbital assignments in parts a and b. The valence *s* orbital matches the A_{1g} representation, so both the *s* and d_{z^2} orbitals of nickel can interact with the zero-node group orbital on the left.

D_{4h}	E	$2C_4$	C_2	$2C_2'$	$2C_2''$	i	$2S_4$	σ_h	$2\sigma_v$	$2\sigma_d$	
Γ	8	0	0	0	0	0	0	0	0	4	
A_{1g}	1	1	1	1	1	1	1	1	1	1	z^2
B_{2g}	1	−1	1	−1	1	1	−1	1	−1	1	xy
A_{2u}	1	1	1	−1	−1	−1	−1	−1	1	1	z
B_{1u}	1	−1	1	1	−1	−1	1	−1	−1	1	
E_g	2	0	−2	0	0	2	0	−2	0	0	(xz, yz)
E_u	2	0	−2	0	0	−2	0	2	0	0	(x, y)

d. Comparisons can be made similarly to those for ferrocene on pages 514–518.

CHAPTER 14: ORGANOMETALLIC REACTIONS AND CATALYSIS

14.1 **a.** $[Mn(CO)_5]^- + H_2C{=}C{-}CH_2{-}Cl \longrightarrow H_2C{=}C{-}CH_2{-}Mn(CO)_4 + Cl^-$

b. Oxidative addition

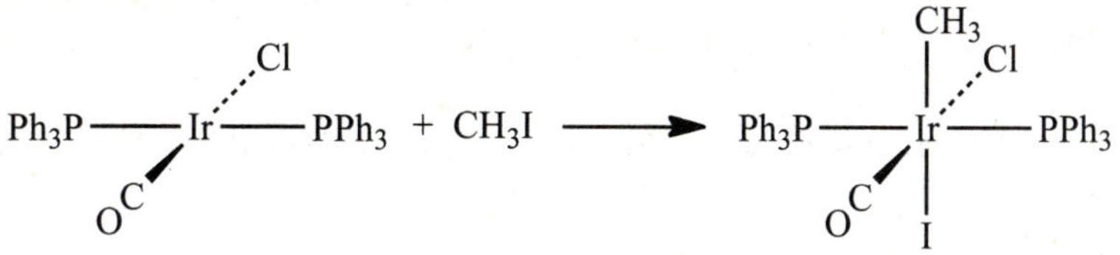

$Ph_3P{-}Ir(Cl)(CO){-}PPh_3 + CH_3I \longrightarrow Ph_3P{-}Ir(CH_3)(Cl)(CO)(I){-}PPh_3$

c. Cyclometallation

$Ph_3P{-}Ir(Cl)(PPh_3){-}PPh_3 \xrightarrow{\Delta}$ (ortho-metallated Ph_2P–C_6H_4–Ir ring) $Ph_2P{-}Ir(Cl)(H)(PPh_3){-}PPh_3$

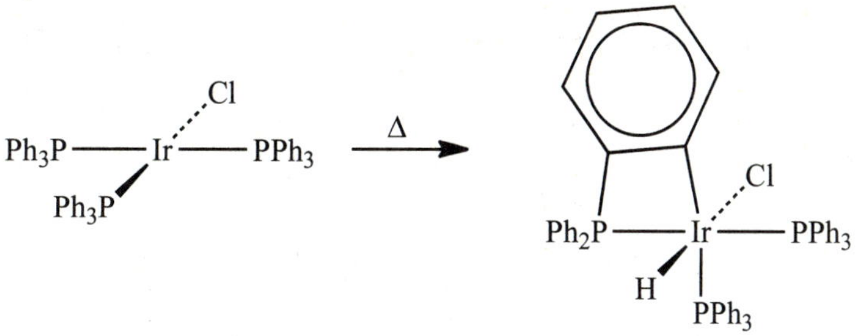

d. Two possibilities: methyl migration and addition, or ligand substitution

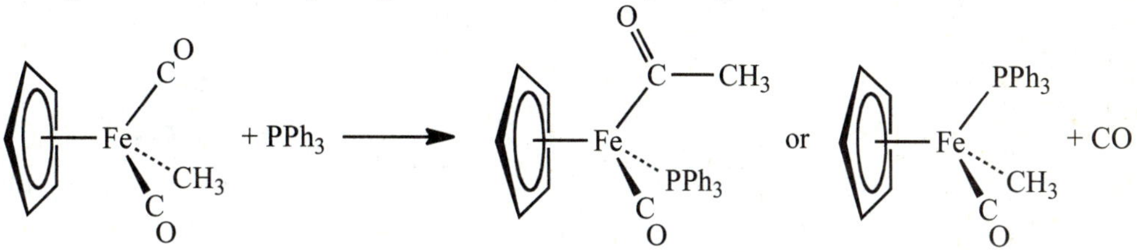

$(\eta^5\text{-}C_5H_5)Fe(CO)_2CH_3 + PPh_3 \longrightarrow (\eta^5\text{-}C_5H_5)Fe(CO)(PPh_3)[C(=O)CH_3]$ or $(\eta^5\text{-}C_5H_5)Fe(CO)(PPh_3)CH_3 + CO$

e. $(\eta^5\text{-}C_5H_5)Mn(CO)_3[C(=O)CH_3] \longrightarrow (\eta^5\text{-}C_5H_5)Mn(CO)_3(CH_3) + CO$
(dissociation and methyl migration)

f. $H_3C{-}Mn(CO)_5 + SO_2 \longrightarrow H_3C{-}S(=O)_2{-}Mn(CO)_5$ (1,1 insertion)

14.2 **a.** $H_3C{-}Mn(CO)_5 + P(CH_3)(C_6H_5)_2 \longrightarrow H_3C(C{=}O){-}Mn(CO)_4[P(CH_3)(C_6H_5)_2]$
(methyl migration and phosphine addition)

b. $[Mn(CO)_5]^- + (\eta^5\text{-}C_5H_5)Fe(CO)_2Br \longrightarrow (CO)_5Mn{-}Fe(\eta^5\text{-}C_5H_5)(CO)_2 + Br^-$
(nucleophilic displacement)

c. *trans*-$Ir(CO)Cl(PPh_3)_2 + H_2 \longrightarrow$ $Ir(H)_2(CO)(Cl)(PPh_3)_2$
(oxidative addition; see page 555)

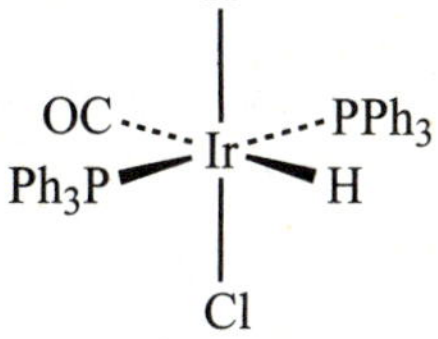

d. $W(CO)_6 + C_6H_5Li \longrightarrow [C_6H_5COW(CO)_5]^- + Li^+$ (nucleophilic attack on carbonyl C; see page 530).

e. Alkyl migration of CH_3 to one of the adjacent CO ligands, followed by addition of ^{13}CO. Isomeric products: 1/3 each of the *fac* enantiomers, 1/3 *mer* isomer. There should be no ^{13}C in the acyl group.

f. Alkyl migration of CH_3 to one of the adjacent CO ligands, followed by addition of ^{13}CO. Isomeric products: two *mer* species, enantiomers if the ^{13}CO is taken into account, identical otherwise.

14.3 **a.** *cis*-$(^{13}CO)(CH_3CO)Mn(CO)_4$ ⟶ $Mn(CH_3)(CO)_5$, 25% with no ^{13}CO, 25% with ^{13}CO *trans* to CH_3, 50% with ^{13}CO *cis* to CH_3.

b. $C_6H_5–CH_2Mn(CO)_5 \xrightarrow{h\nu} CO +$ [structure: η³-benzyl Mn(CO)₄]

c. $[V(CO)_6] + NO \longrightarrow [V(CO)_5(NO)] + CO$

d. $Cr(CO)_6 + 2\ Na/NH_3 \longrightarrow 2\ Na^+ + [Cr(CO)_5]^{2-} + CO$

e. $Fe(CO)_5 + NaC_5H_5 \longrightarrow Na^+ + [(\eta^5\text{-}C_5H_5)Fe(CO)_2]^- + 3\ CO$

f. $[Fe(CO)_4]^{2-} + CH_3I \longrightarrow [(CH_3)Fe(CO)_4]^- + I^-$

g. $(Ph_3P)_3RhCH_3 \xrightarrow{\Delta} CH_4 +$ [structure: ortho-metallated Ph_2P–C_6H_4 bound to Rh, with PPh_3 and Ph_3P ligands]

14.4 $[(C_5H_5)Fe(CO)_3]^+ + NaH \rightarrow$ **A** $(C_7H_6O_2Fe)$ **A** $= (C_5H_5)Fe(CO)_2H$

A ⟶ **B** (colorless gas) + **C** $(C_7H_5O_2Fe)$ (purple-brown)   **B** $= H_2$

(See Figure 13.35, page 522 for structure of **C**.)   **C**$= [(C_5H_5)Fe(CO)_2]_2$

C $+ I_2$ ⟶ **D** $(C_7H_5O_2FeI)$ (brown)   **D** $= (C_5H_5)Fe(CO)_2I$

D $+ TlC_5H_5$ ⟶ **E** $(C_{12}H_{10}O_2Fe) + TlI$ **E** $= (C_5H_5)_2Fe(CO)_2$, one Cp η^1
(See Figure 13.35, page 522 for structure of **E**.)

E ⟶ **F** $(C_{10}H_{10}Fe)$ + colorless gas (CO)   **F** $= (C_5H_5)Fe$, ferocene

14.5 $[(\eta^5\text{-}C_5H_5)Fe(CO)_2]^- + ClCH_2CH_2SCH_3 \longrightarrow$ **A** $(C_{10}H_{12}FeO_2S)$

A + heat $\longrightarrow$ **B**

A has bands at 1980 and 1940 cm^{-1}, **B** at 1920 and 1630 cm^{-1}.

The sulfur reagent loses Cl^- and bonds to the Fe as an alkyl ligand to form **A**, with two carbonyls. This is an example of nucleophilic displacement of chloride by an organometallic anion. **A** then rearranges, with the S becoming attached to Fe, and the alkyl migrates to a carbonyl carbon. **B** contains an ordinary carbonyl and an acyl C=O bond, for the two quite different C–O stretching energies.

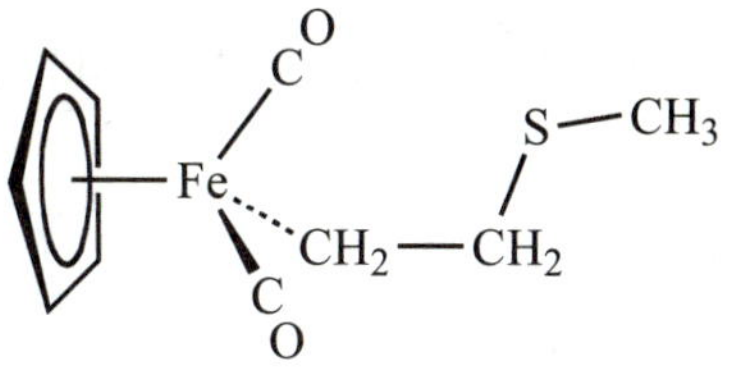

O C Fe C O CH$_2$—CH$_2$ S—CH$_3$

A

O C Fe S CH$_3$ CH$_2$ C O CH$_2$

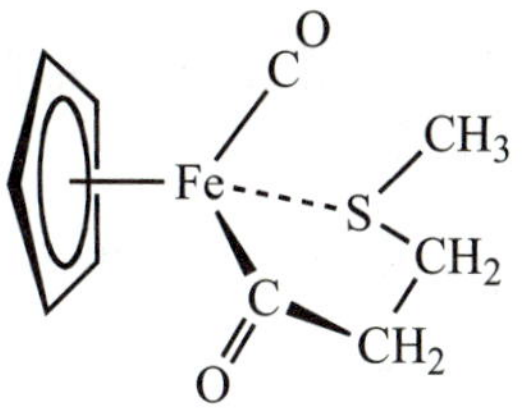

B

14.6 **a.** Two term rate laws like this could be the result of two parallel associative reactions, the first by solvent, the second by the phosphite, or could result from a dissociative reaction for the first term and an associative reaction for the second.

First term, dissociative, k_1 $[V(CO)_5(NO)]$:

$$V(CO)_5(NO) \longrightarrow V(CO)_4(NO) + CO \xrightarrow{PR_3} V(CO)_4(NO)(PR_3)$$

Second term, associative, $k_2[PR_3][V(CO)_5(NO)]$:

$$V(CO)_5(NO) \xrightarrow{PR_3} V(CO)_5(NO)(PR_3) \longrightarrow V(CO)_4(NO)(PR_3) + CO$$

b. $V(CO)_5[P(OCH_3)_3](NO)$ If the NO is a bent, 1-electron donor, it can be an 18-electron species ($5 + 5\times2 + 2 + 1 = 18$). If NO is a linear, 3-electron donor, the total is 20 electrons.

14.7 $Co_2(CO)_8 \rightleftharpoons Co_2(CO)_7 + CO$ K_1 (fast equilibrium) $[Co_2(CO)_7] = K_1[Co_2(CO)_8]/[CO]$

$Co_2(CO)_7 + H_2 \longrightarrow Co_2(CO)_7H_2$ k_2 (slow)

$Co_2(CO)_7H_2 + CO \longrightarrow 2\ HCo(CO)_4$ (fast)

Rate $= k_2[Co_2(CO)_7][H_2] = k_2K_1\ [Co_2(CO)_8][H_2]/[CO]$

14.8 This depends on the cone angle of the phosphine ligands, with the order $PPh_3 > PBu_3$ (estimated) $> P(OPh)_3 > P(OMe)_3$ from Table 14.1 (page 554). The PPh_3 should dissociate most rapidly and the $P(OMe)_3$ should dissociate least rapidly.

14.9 K for the dissociation reaction is in the order $PPh_3 > PMePh_2 > PEt_3 > PMe_3$, as a result of a combination of decreasing cone angle and increasing negative charge on the phosphorus. Alkyls push more electron density onto P than phenyl rings.

14.10 If it loses CO followed by migration of CH_3 to an adjacent position, all the CO lost should be ^{12}CO, because the CO ligands *cis* to the $CH_3C{=}O$ will be the ones lost.

14.11 **a.** There are two possible products, both a result of methyl migration followed by carbonyl addition. The methyl group can move to either the 1 position or the 2 position. The resulting products are different only in the location of the ^{13}CO.

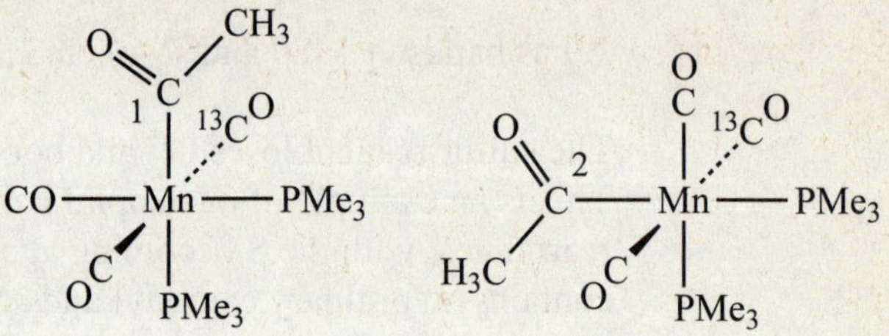

b. The new IR band is from the acyl carbonyl, and should be near 1630 cm^{-1} (see problem 14.5).

14.12 $[(C_5H_5)_2Fe_2(CO)_4]$ + Na/Hg ⟶ **A**

A + Br_2 ⟶ **B**

B + $LiAlH_4$ ⟶ **C**

C + PhNa ⟶ **A** + **D**

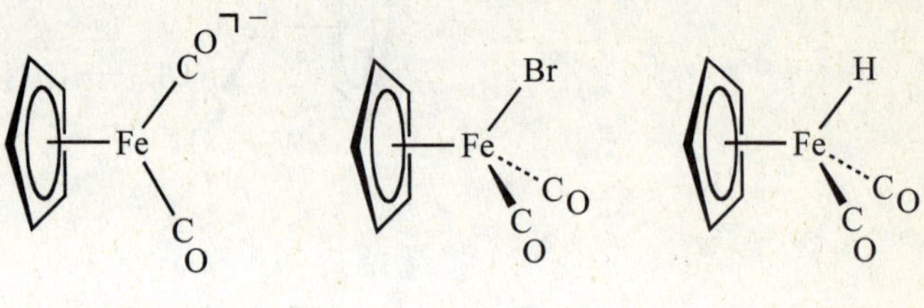

A **B** **C**

A: Sodium acts as reducing agent. Anion **A** has lower energy C–O stretches, as expected (see page 498).

B: Br_2 acts as oxidizing agent and is the source of the bromo ligand.

C: The NMR peak at –12 ppm is caused by the hydride ligand.

D: benzene, C_6H_6

14.13 **a.** The bromoethoxide anion adds to a carbonyl; the hard oxygen of the anion adds to the hard carbon. Then Br^- is lost, leaving a positive carbon. The alkyl tail can then bend around and react with the carbonyl oxygen, giving the compound shown below.

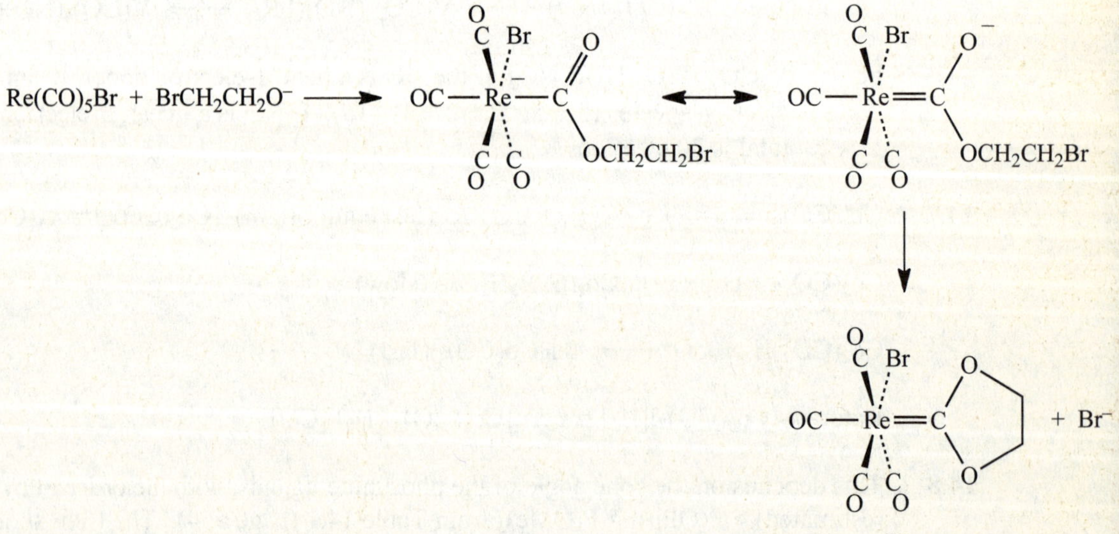

$Re(CO)_5Br + BrCH_2CH_2O^-$ ⟶

Y + Br^-

b. There are two electrons donated from each of the six ligand positions, and Re^+ has 6 electrons for a total of 18. In the isomer shown there are three different carbonyls, and the carbene ligand has two identical carbons, so there are five different magnetic environments for carbon. In the other possible isomer, with Br trans to the alkoxide ligand, there are not five different carbon environments. Finally, Ag^+ can remove Br^- from the complex.

14.14 a. This is the example described on pages 543–544. See those pages for further details.

I + PPh_3 ⟶ **II** **II:** IR: 2038, 1958, 1906 cm^{-1}
NMR: 7.62, 7.41 muliplets (15), 4.19 multiplet (4)

II + PPh_3 ⟶ **III** **III:** IR: 1944, 1860 cm^{-1}
NMR: 7.70, 7.32 multiplets (15), 3.39 singlet (2)

II has 3 CO ligands in a *mer* geometry (see Table 13.7, page 537), so the CO *trans* to Br has been replaced by PPh_3. The NMR shows one PPh_3 (15) and the ethylene hydrogens (4).

III has only 2 CO ligands in a *cis* configuration (Table 13.7), so an axial CO has been replaced. The ratio of NMR integrated peaks is now 15:2 because there are two PPh_3 groups.

b. **I** + $Ph_2PCH_2PPh_2$ ⟶ **IV**

IV 2036, 1959, 1914 cm^{-1}, 35.8% C, 2.73% H

Replacing two CO's with the phosphine gives a compound with 25.7% C, 3.1% H. Replacing one CO and the Br with the phosphine gives a compound with 28.9% C, 3.3% H.

The only way to get the analysis to work out is to have a single diphos ligand bridging two Re atoms after loss of CO from each:

$[(Re(CO)_3Br(C(O_2C_2H_4))]_2\{(PPh_2)_2CH_2\}$, **IV**, has 36.14% C, 2.46% H.

c. **I** + $S_2CN(CH_3)_2^-$ ⟶ $Re(CO)_5Br$ + **V**

V has no metal, has no IR bands between 1700 and 2300 cm^{-1}, has NMR bands at 3.91 (triplet), 3.60 (triplet), 3.57 (singlet), and 3.41 (singlet).

Acting as a nucleophile, the dithiocarbamate ion attacks the carbene to form $(CH_3)_2NC(=S)SC_2H_4O^-$, which can pick up a proton from trace amounts of water to make the hydroxy compound $(CH_3)_2NC(=S)SC_2H_4OH$, **V**. The 1500 and 977 cm^{-1} bands are the N–C and C–S bands, and the NMR shows triplets for the two CH_2 units, singlets for the CH_3's.

14.15 The mechanism is the one described in problem 13, with formation of the alkoxide ion by the reaction:

$$\triangleright O + Br^- \longrightarrow BrCH_2CH_2O^-$$

14.16 The reaction proceeds by substitution of $Mn(CO)_5^-$ for Br^-, followed by alkyl migration, addition of $Mn(CO)_5^-$, and finally cyclization:

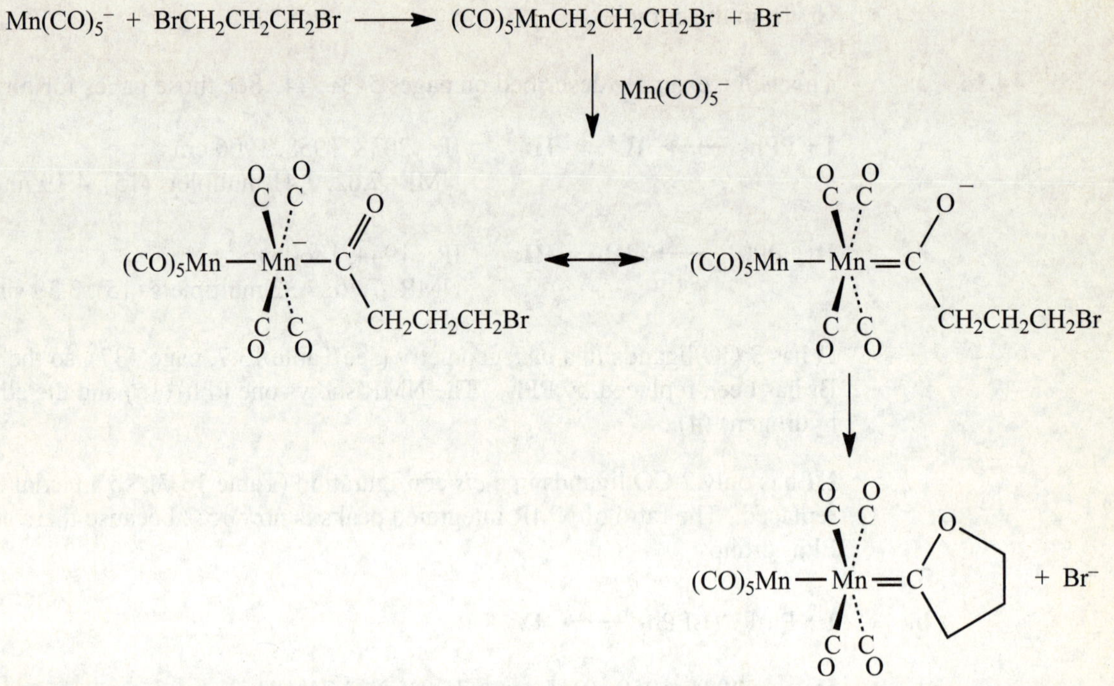

$$Mn(CO)_5^- + BrCH_2CH_2CH_2Br \longrightarrow (CO)_5MnCH_2CH_2CH_2Br + Br^-$$

$\downarrow Mn(CO)_5^-$

$(CO)_5Mn—Mn^-(CO)_4—C(=O)CH_2CH_2CH_2Br \longleftrightarrow (CO)_5Mn—Mn(CO)_4{=}C(O^-)CH_2CH_2CH_2Br$

$\downarrow$

$(CO)_5Mn—Mn(CO)_4{=}C$ (cyclic, with ring O) $+ Br^-$

14.17 The acyl metal carbonyl has a resonance structure with a negative charge on oxygen, where the proton can add. Protonation is then similar to the alkylation reaction described in Chapter 13 (page 530).

14.18 **a.** Acetaldehyde from ethylene:

$$C_2H_4 + CO + H_2 \xrightarrow{[PdCl_4]^{2-}} H_3C—C(=O)H + H_2O$$

The Wacker process (Figure 14.17, page 573) with ethylene as the starting alkene will result in acetaldehyde.

b. Ethyl propionate from chloroethane:

$$[Co(CO)_4]^- \xrightarrow{CH_2CH_2Cl} CH_3CH_2Co(CO)_4 \xrightarrow{CO} (CH_3CH_2CO)Co(CO)_4 \xrightarrow{C_2H_5OH} CH_3CH_2COOC_2H_5 + HCo(CO)_4$$

c. Pentanal from 1-butene: The hydroformylation process (Figure 14.14, page 568) does this.

d. 4-phenylbutanal from an alkene: Again, the hydroformylation process should do this, starting with 4-phenyl-1-propene.

e. Wilkinson's catalyst $[ClRh(PPh_3)_3]$ (page 574) should do this. D_2 would be added across the least hindered double bond.

f. Catalytic deuteration with H_3TaCp_2 (Figure 14.13, page 567) should deuterate the phenyl ring without affecting the methyl hydrogens.

14.19 Figure 14.15, page 570 shows this process of hydroformylation. *n*-pentanal results if $R = C_2H_5$. Identification of the steps:

1. Dissociation of one CO
2. Addition of the alkene
3. 1,2 insertion
4. Addition of CO
5. Alkyl migration
6. Oxidative addition
7. Reductive elimination of the aldehyde

14.20 Hydroformylation (Figure 14.15, page 570) with $Rh(CO)_2(PPh_3)_2$ as the catalytic species will work, starting with 2-methyl-1-butene: $H_3C–CH_2–CH(CH_3)=CH_2$.

14.21 **a.** Direct metathesis would occur as follows:

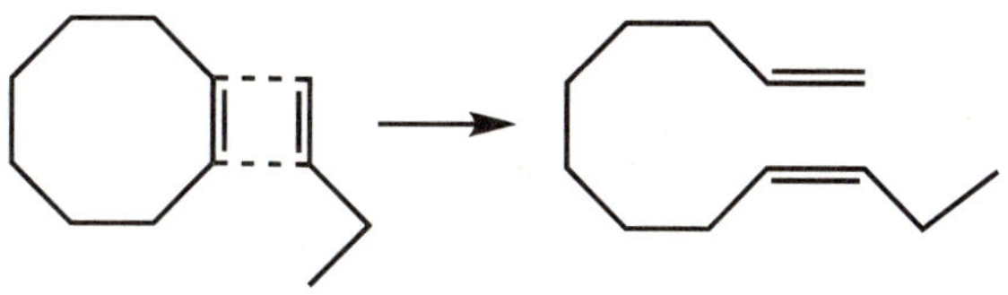

In addition, reactants could undergo self-metathesis, for example

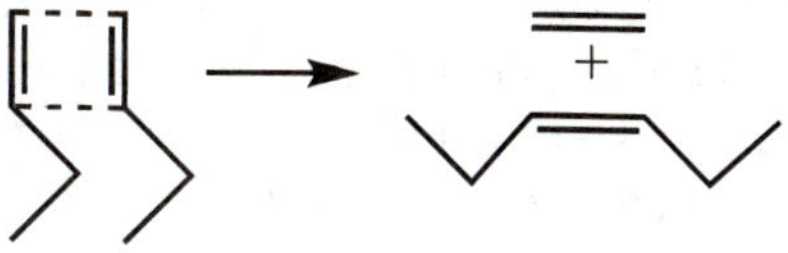

+

and the products of the direct metathesis could undergo further metathesis to give a variety of products. One of these could be formed as shown here:

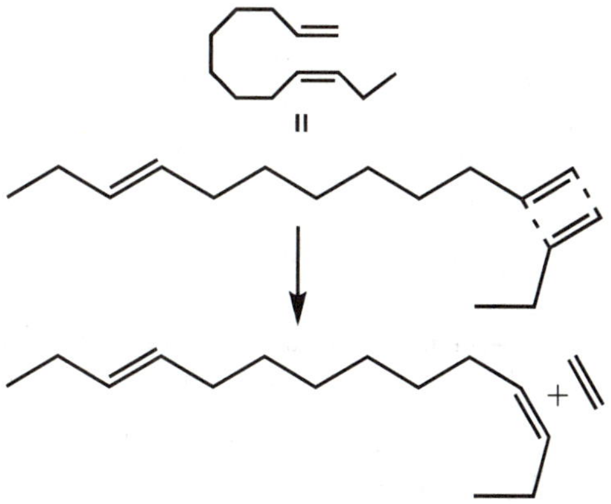

II

+

b. This is Hérisson and Chauvin's classic experiment, with products shown in Figure 14.23. As in part a, self metathesis can also occur, and the products of direct metathesis can also undergo further metathesis to give a variety of products.

c.

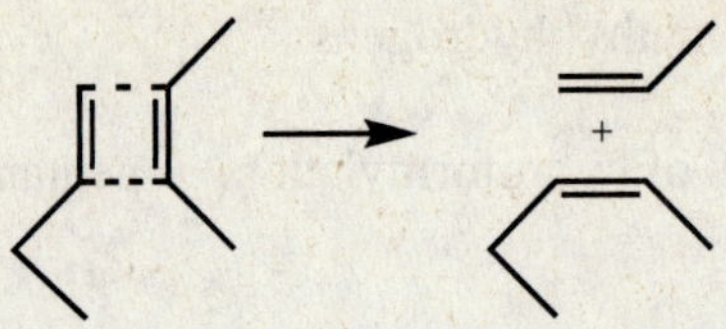

d.

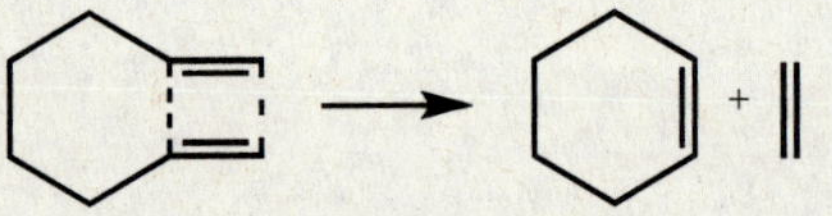

14.22 a.

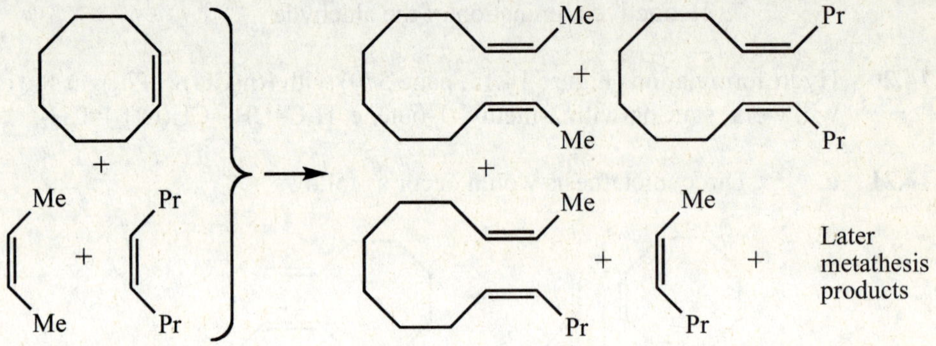

Me

Pr

Me

Pr

Me

Pr

Me

Pr

Me

Me

Later metathesis products

Pr

Pr

"Double cross" product

Me = methyl
Pr = *n*-propyl

b. The pairwise mechanism would call for the dimethyl and dipropyl products to be formed first, followed by the "double cross" product. In the non-pairwise mechanism, the double cross product should form simultaneously with the dimethyl and the dipropyl products. Experimental results showed that the double cross product formed (along with the dimethyl and dipropyl products) at the beginning of the reaction, consistent with the non-pairwise mechanism.

14.23 a.

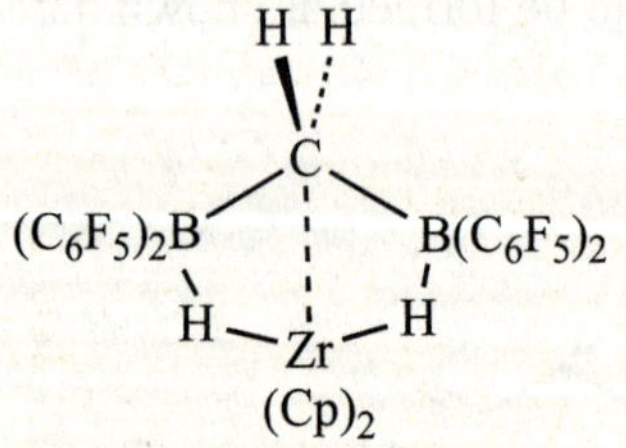

H H
C
$(C_6F_5)_2B$ $B(C_6F_5)_2$
H Zr H
$(Cp)_2$

b. One possible route:

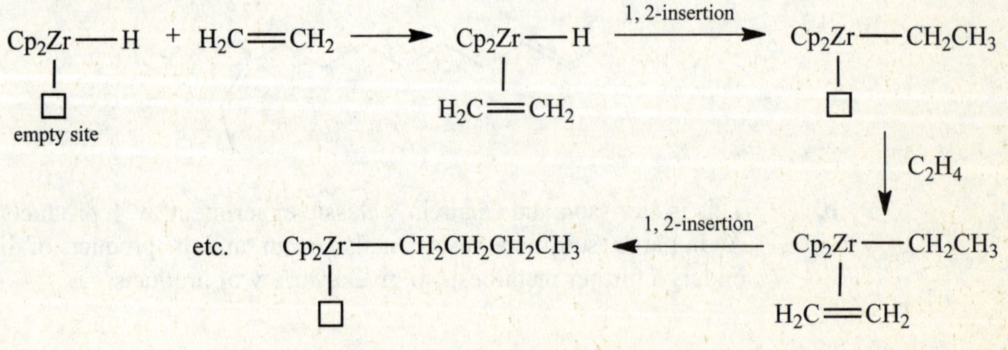

$Cp_2Zr—H$ + $H_2C{=}CH_2$ → $Cp_2Zr—H$ ($H_2C{=}CH_2$) —1, 2-insertion→ $Cp_2Zr—CH_2CH_3$ (□)

empty site

—C_2H_4→ $Cp_2Zr—CH_2CH_3$ ($H_2C{=}CH_2$) —1, 2-insertion→ $Cp_2Zr—CH_2CH_2CH_2CH_3$ (□) etc.

14.24

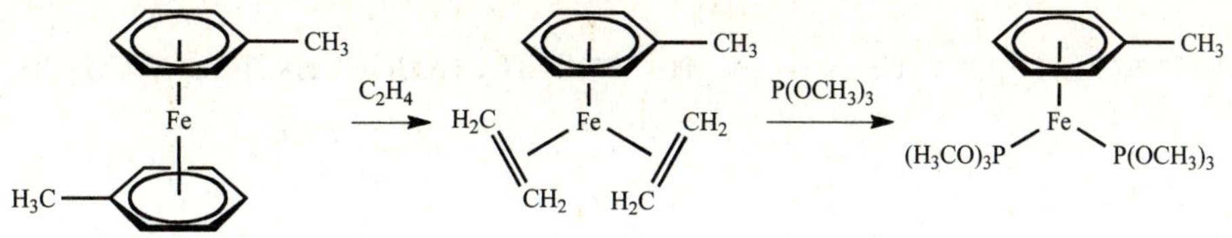

CH_3 — Fe — H_3C $\xrightarrow{C_2H_4}$ CH_3 — Fe (H_2C, CH_2, CH_2, H_2C) $\xrightarrow{P(OCH_3)_3}$ CH_3 — Fe ($(H_3CO)_3P$, $P(OCH_3)_3$)

X

14.25 $RhCl_3 \cdot 3\ H_2O + P(o\text{-}MePh)_3 \longrightarrow$ **I** (blue-green) ($C_{42}H_{42}P_2Cl_2Rh$) $\mu_{eff} = 2.3$ B.M., Rh–Cl $\nu = 351\ cm^{-1}$

I is *trans*-$RhCl_2(PR_3)_2$ The IR is the Rh–Cl asymmetric stretch. The *cis* isomer would give two IR bands. One unpaired electron (15 electron species, square planar).

I + heat $\longrightarrow$ **II** (yellow, diamagnetic) Rh:Cl = 1, $\nu = 920\ cm^{-1}$

II Loss of one Cl and combination of two CH_3's to form a single tridentate phosphine ligand with a π bond. Sixteen electrons around Rh. The double bond in the ligand is perpendicular to the Rh–Cl–P plane because of the size and geometry of the benzene rings.

II + SCN^- $\longrightarrow$ **III** Rh(SCN)

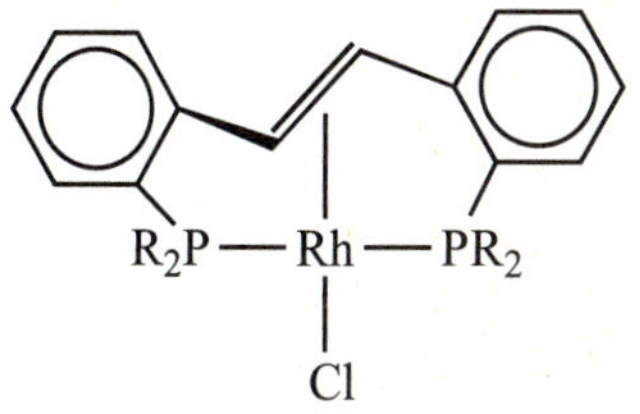

NMR: δ	Area	Type
6.9 – 7.5	12	aromatic
3.50	1	doublet of 1:2:1 triplets
2.84	3	singlet
2.40	3	singlet

II:

R_2P — Rh — PR_2

Cl

III: Same overall structure as **II**, with SCN substituted for Cl. The singlets are the methyl protons (two separate environments—not obvious from the drawing), the doublet is from the vinylic protons, and the aromatic multiplets are the sum of all the phenyl protons.

II + NaCN $\longrightarrow$ **IV** ($C_{21}H_{19}P$, mw = 604) $\nu = 965\ cm^{-1}$

NMR: δ	Area	Type
7.64	1	singlet
6.9 - 7.5	12	aromatic
2.37	6	singlet

IV is the diphosphine ligand, $C_{42}H_{38}P_2$, shown above in compound **II**. (CN^- substitutes for all the ligands in this reaction.) The IR band is characteristic of *trans* vinylic hydrogens. All methyl protons are equivalent in the free ligand (singlet), the vinylic protons are a singlet at 7.64, and the phenyl protons are a multiplet at 6.9–7.5. The change in the vinyl hydrogen IR band shows that the ethylene does coordinate to Rh. Coordination reduces the C=C bonding, which also reduces the C–H bending energy (electrons are drawn away from C–H, toward C–Rh).

14.26 **a.**

$[MeC{\equiv}C(CH_2)_2OOC(CH_2)_2]_2 = H_3C{\equiv}CCH_2CH_2OOCCH_2CH_2COOCH_2CH_2C{\equiv}CCH_3$

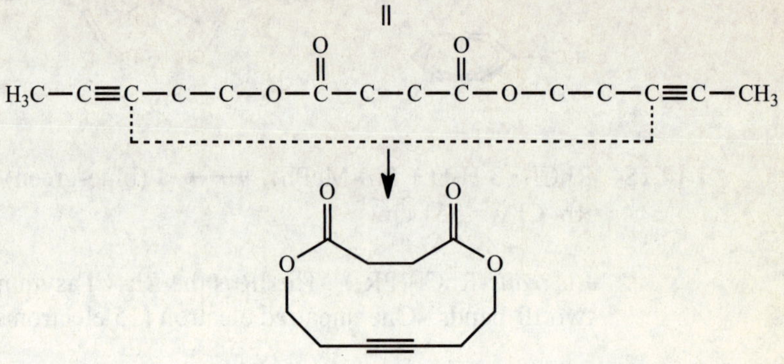

$H_3C{-}C{\equiv}C{-}C{-}C{-}O{-}C(=O){-}C{-}C{-}C(=O){-}O{-}C{-}C{-}C{\equiv}C{-}CH_3$

b.

$MeC{\equiv}C(CH_2)_8COO(CH_2)_9C{\equiv}CMe$

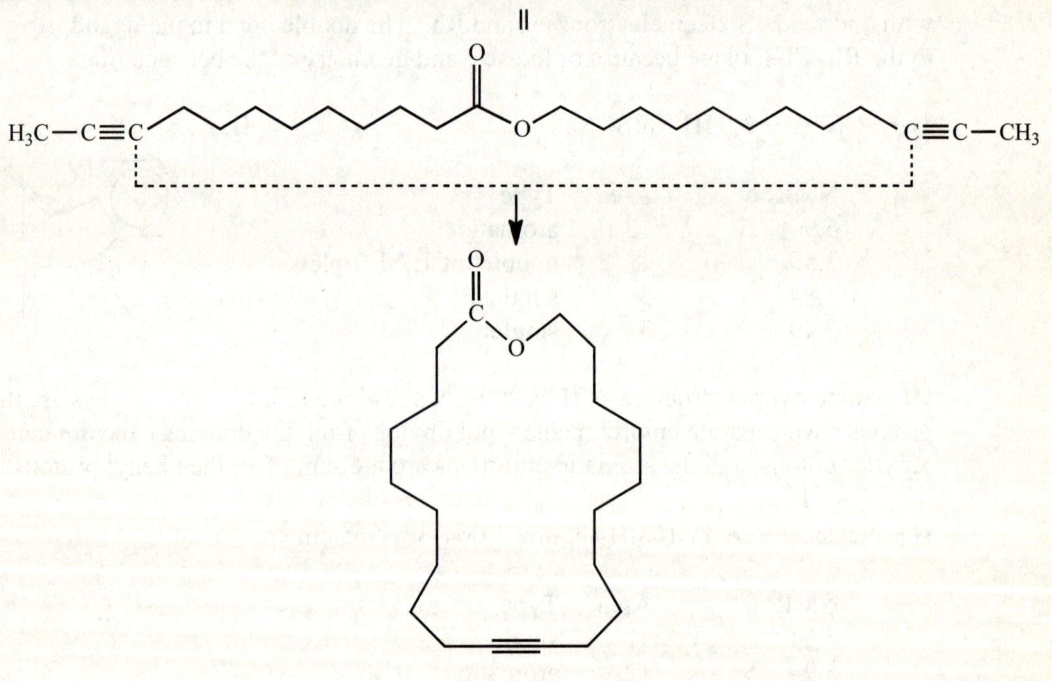

$H_3C{-}C{\equiv}C \ldots C{\equiv}C{-}CH_3$

14.27 Evidence in support of the intermediate shown at the right includes:

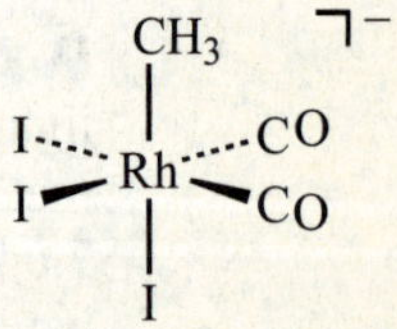

$[CH_3Rh(CO)_2I_3]^-$

(1) An infrared band at 2104 cm^{-1} has been observed. This band is similar to the higher energy band in the Ir analogue, $[CH_3Ir(CO)_2I_3]^-$, which has carbonyl bands at 2102 and 2049 cm^{-1}. A second band, expected for $[CH_3Ir(CO)_2I_3]^-$, would be hidden under strong bands of the reactants and products of steps 2 and 3 of the mechanism.

(2) The ratio of the absorbance of the 2104 cm^{-1} band to the absorbance of a band at 1985 cm^{-1} of $[Rh(CO)_2I_2]^-$, the reactant in step 2, is proportional to the concentration of CH_3I. This is consistent with what would be expected

from the equilibrium constant expression for formation of the intermediate,

$$K = \frac{[Rh(CO)_2I_2]^-[CH_3I]}{[CH_3Rh(CO)_2I_3]^-}$$

and is consistent with the steady state approximation for the mechanism.

(3) The maximum intensity of the 2104 cm^{-1} band occurs when the product of step 3 is formed most rapidly.

(4) When $^{13}CH_3I$ is used, a doublet is observed in the NMR consistent with ^{13}C–^{103}Rh coupling. Other NMR data also support the proposed structure.

14.28 **A**: $[Fe(CO)_3(CN)_3]^-$ **B**: *cis*-$[Fe(CO)_2(CN)_4]^-$

The cyanide ion has the capacity to replace ligands such as I^- and CO. The C–N stretch involves a large change in dipole moment, so this vibration, like that of CO, can be useful in characterizing cyano complexes. Because CN has a slightly smaller reduced mass than CO, C–N stretches typically occur at slightly higher energies than C–O stretches (see page 498).

In this situation, both **A** and **B** have two sets of C–N and C–O stretches, indicating that both complexes have at least two of both ligands. As more cyano ligands are added, the concentration of electrons on Fe increases, and both types of ligands become stronger π acceptors, reducing the energy of the stretching vibrations.

The formula of **B** suggests the possibility of both *cis* and *trans* isomers. The observation of two bands in the carbonyl region is consistent with the *cis* isomer. (See also problem 13.23 and its reference.)

14.29 **2**:

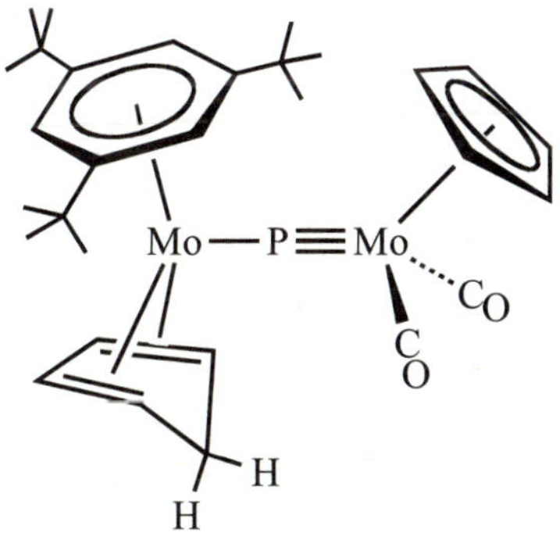

The two IR bands are as expected for this dicarbonyl complex. NMR peaks can be assigned as follows: chemical shift 5.28 (relative area 5): Cp; 1.31 (27): nine methyl groups on *t*-butyl groups on benzene ring; 5.15 (3): protons on benzene ring; 5.46 (2), 4.22 (2) and hidden small peak: η^4-C_5H_6. In the ^{13}C NMR, the resonance at 236.9 ppm can be assigned to the carbonyl carbons. This product is formed via an unusual attack of a hydride on a cyclopentadienyl carbon. For an additional example of this type of attack, see footnote 17 in the reference.

14.30 **X:** 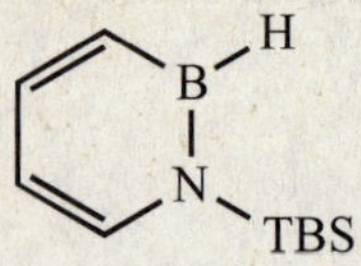 **Y:** 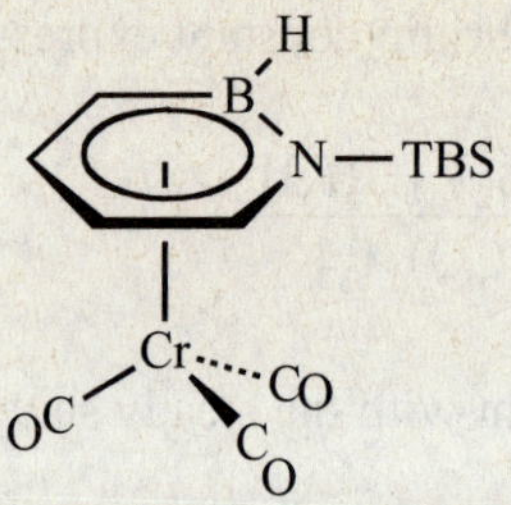 **Z:**

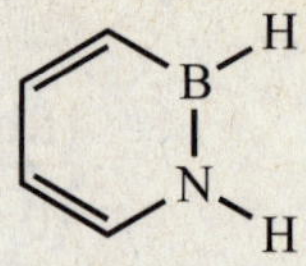

In heterocycle **Z** coupling of the N–H proton with the ^{14}N nucleus (which has $S = 1$) results in a triplet, and coupling for the B–H proton with the ^{11}B nucleus (which has $S = 3/2$) results in a quartet. (The original spectrum is shown in the reference.)

14.31 See Figure 13.32 for a diagram of this molecular "ferrous wheel"!

CHAPTER 15: PARALLELS BETWEEN MAIN GROUP AND ORGANOMETALLIC CHEMISTRY

15.1 **a.** $Mn_2(CO)_{10} + Br_2 \longrightarrow 2\ Mn(CO)_5Br$

b. $HCCl_3 + \text{excess}[Co(CO)_4]^- \longrightarrow$

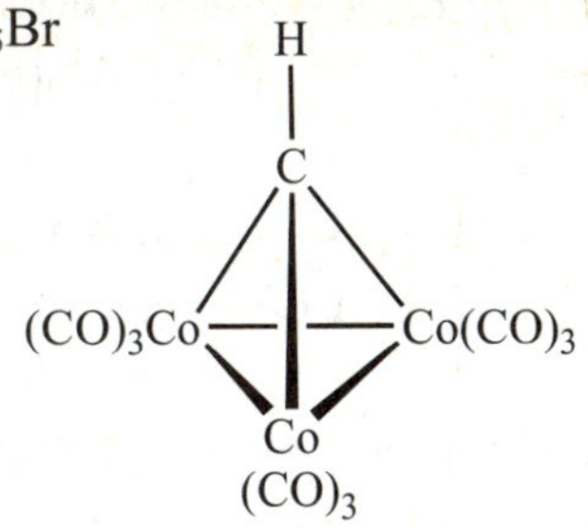

c. $Co_2(CO)_8 + (SCN)_2 \longrightarrow 2\ NCSCo(CO)_4$

d. $Co_2(CO)_8 + C_6H_5–C{\equiv}C–C_6H_5 \longrightarrow$

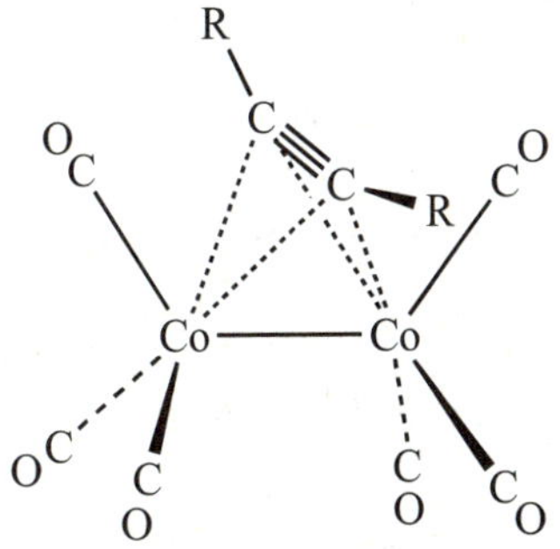

e. $Mn_2(CO)_{10} + [(\eta^5\text{-}C_5H_5)Fe(CO)_2]_2 \longrightarrow 2\ (CO)_5Mn–Fe(\eta^5\text{-}C_5H_5)(CO)_2$

15.2 **a.** $Tc(CO)_5 \underset{\circ}{\longleftrightarrow} CH_3$

b. $[Re(CO)_4]^- \underset{\circ}{\longleftrightarrow} CH_2$

c. $[Co(CN)_5]^{3-} \underset{\circ}{\longleftrightarrow} CH_3$

d. $[CpFe(C_6H_5)]^+ \underset{\circ}{\longleftrightarrow} CH_4$

e. $[Mn(CO)_5]^+ \underset{\circ}{\longleftrightarrow} CH_3^+$

f. $Os_2(CO)_8 \underset{\circ}{\longleftrightarrow} H_2C{=}CH_2$

15.3 Many examples are possible. Two for each:

a. $CH_3 \underset{\circ}{\longleftrightarrow} (\eta^6\text{-}C_6H_6)Mn(PPh_3)_2$
$CH_3 \underset{\circ}{\longleftrightarrow} [Ni(CS)(PMe_3)_2Cl_2]^+$

b. $CH \underset{\circ}{\longleftrightarrow} (\eta^4\text{-}C_4H_4)Co(CS)$
$CH \underset{\circ}{\longleftrightarrow} (\eta^6\text{-}C_6H_6)Rh$

c. CH_3^+ ⟵o⟶ Mo(borazine)(PMe_3)(CS)
CH_3^+ ⟵o⟶ $[V(CO)_3(en)]^-$

d. CH_3^- ⟵o⟶ (η^8-C_8H_8)Ru(PEt_3)
CH_3^- ⟵o⟶ $[Co(N_2)(CO)_2(bipy)]^+$

e. (η^5-C_5H_5)$Fe(CO)_2$ ⟵o⟶ $Co(CO)_3Cl_2$
(η^5-C_5H_5)$Fe(CO)_2$ ⟵o⟶ $[(\eta^6\text{-}C_6H_6)Cr(CO)_2]^-$

f. $Sn(CH_3)_2$ ⟵o⟶ (η^5-C_5H_5)Ir(CO)
$Sn(CH_3)_2$ ⟵o⟶ $[Tc(CO)_4]^-$

15.4 Many possibilities exist; some of those given here may not have been made, and may not be stable configurations.

a. C_2H_4 ⟵o⟶ $(CO)_4Fe{=}Fe(CO)_4$

b. P_4 ⟵o⟶ $[Ir(CO)_3]_4$

c. *cyclo*-C_4H_8 ⟵o⟶

$(CO)_4Fe$ — $Ru(CO)_4$
| |
$(CO)_4Ru$ — $Fe(CO)_4$

d. S_8 ⟵o⟶ *cyclo*-$[Fe(CO)_4]_8$

15.5 a. All three complexes are 18 electron species, with the benzene ring of $[(C_5Me_5)Fe(C_6H_6)]^+$ replaced by three carbonyl groups or two carbonyls and a phosphine. All have a formal coordination number of six.

b. The experimental results are more complex than might be expected.

$[(C_5Me_5)Fe(C_6H_5)]^+ + H^-$ ⟶ Fe H H

$[(\eta^5\text{-}C_5H_5)Fe(CO)_3]^+ + H^-$ ⟶ $(\eta^5\text{-}C_5H_5)Fe(CO)_2H$ ⟶ $[(\eta^5\text{-}C_5H_5)Fe(CO)_2]_2$

$[(\eta^5\text{-}C_5H_5)Fe(CO)_2PPh_3]^+ + H^-$ ⟶ PPh_3 OC CO Fe H H

(See A. Davison, M. L. H. Green, and G. Wilkinson, *J. Chem. Soc.*, **1961**, 3172.)

15.6 **a.** CH_2, $Fe(CO)_4$, $[Mn(CO)_4]^-$, and PR_2 each has two frontier orbitals, each with one electron. Each fragment is two ligands short of the parent polyhedron (octahedron or tetrahedron).

b. $Fe(CO)_4$ and CpRh(CO) each has two frontier orbitals, each with one electron.

c. $[Re(CO)_4]^-$ has two frontier orbitals, with one electron in each. R_2C is isolobal with two frontier orbitals, each with one electron.

15.7 The $W(Cp)(CO)_2$ fragments are isolobal with CR and CPh, each with three orbitals containing one electron each; PtR_2 and $Cu(C_5Me_5)$ are isolobal, each with two orbitals containing one electron each.

15.8 **a.** If Mn has the lower energy orbital, then the bonding molecular orbital has a greater contribution by the Mn orbital, and the electrons in the orbital are polarized toward Mn.

b. The gold orbitals are higher in energy. Rather than matching energies with the lowest of the π orbitals of the Cp ring, the Au orbitals will match better with the higher π orbitals, which have a nodal plane cutting across the ring.

15.9 **a.** CH_2 and $Fe(CO)_4$ are isolobal, each with two orbitals containing one electron apiece.

b. $Mn(CO)_2(C_5Me_5)$ is isolobal with $[Mn(CO)_5]^+$, a 16 electron species, which is in turn isolobal with CH_2. The Mn–Sn–Mo fragment is similar to allene, C=C=C, in which the double bonds force a linear geometry.

15.10 $[C(AuPPh_3)_5]^+$ Structure: slightly distorted trigonal bipyramid, as expected from VSEPR.

$[C(AuPPh_3)_6]^{2+}$ Structure: slightly distorted octahedron, as expected from VSEPR.

In these complexes, each of the $AuPPh_3$ groups can be viewed as having an *sp* hybrid orbital pointing in toward the carbon. Interactions between these hybrids with the *s* and *p* orbitals of carbon give rise to four bonding orbitals in each case. The eight valence electrons available fill these orbitals. The result is the equivalent of four bonds spread over the complex (bond order of 4/5 in $[C(AuPPh_3)_5]^+$ and 4/6 in $[C(AuPPh_3)_6]^{2+}$. Metal–metal bonding is also likely to contribute to the stability of these structures.

For more details, see H. Schmidbauer, et. Al., *Angew. Chem. Int. Ed. Engl.*, **1989**, *28*, 463 and *Angew. Chem. Int. Ed. Engl.*, **1988**, *27*, 1544.

15.11 Carbonyl ligands are significantly stronger π acceptors than cyclopentadienyl. Replacing three CO ligands with Cp on each metal atom, therefore, leaves the metals with greater electron density for involvement in metal–metal bonding, leading to shorter metal–metal bonds in the cobalt complexes. In addition, there is a general decrease in atomic radii from left to right in this row of transition metals, so Co atoms are slightly smaller than Fe atoms (see Table 2.8).

15.12 **a.** The orbital interactions are similar in the two cases, except that the π orbitals of the P_5 ring are lower in energy than those of the C_5H_5 ring, resulting in a stronger ability of the P_5 ring to accept electrons from the metal (see also reference 11 in this chapter).

b. Because the P_5 ring acts as a strong acceptor, electrons are accepted into antibonding orbitals in the titanium complexes, leading to longer P–P distances than in P_5^- (calculated at 2.12 Å). The Ti–P distances are calculated to be shorter in the theoretical structure $[Ti(P_5)]^-$ than in $[Ti(P_5)_2]^{2-}$, in which ligands compete for electrons in the same *d* orbitals on the metal. The consequence is that the predicted P–P distance is longer (2.24 Å) in $[Ti(P_5)]^-$, where the ligand can act as a stronger acceptor because it is closer to the metal, than in $[Ti(P_5)_2]^{2-}$ (calculated: 2.175 Å; experimental in reference 11: 2.154 Å).

15.13 a. The $Mn(CO)_5$ fragments have a single electron in their HOMO, largely derived from the d_{z^2} (hybridized with the p_z orbital) of Mn. Lobes of the HOMOs of two $Mn(CO)_5$ fragments interact in a sigma fashion with the σ_g orbital of C_2 (see Figure 5.5, page 133 for the approximate shape) derived primarily from the p_z orbitals of the C atoms:

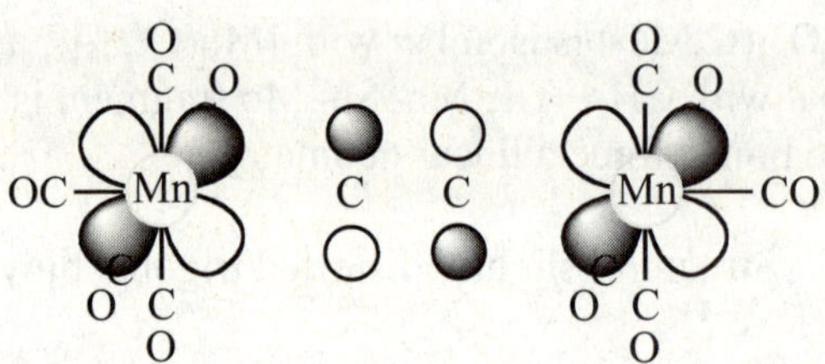

b. The empty π^* orbitals of C_2 can interact with occupied *d* orbitals of $Mn(CO)_5$ fragments:

c. The reference points out other interactions, such as between π (bonding) orbitals of C_2 and $Mn(CO)_5$, and discusses the relative energies of the molecular orbitals of this molecule and other molecules having bridging C_2 ligands.

15.14 A staggered configuration is more likely, as predicted by VSEPR. The bonding is similar to that in the triply bonded $[Os_2Cl_8]^{2-}$.

15.15

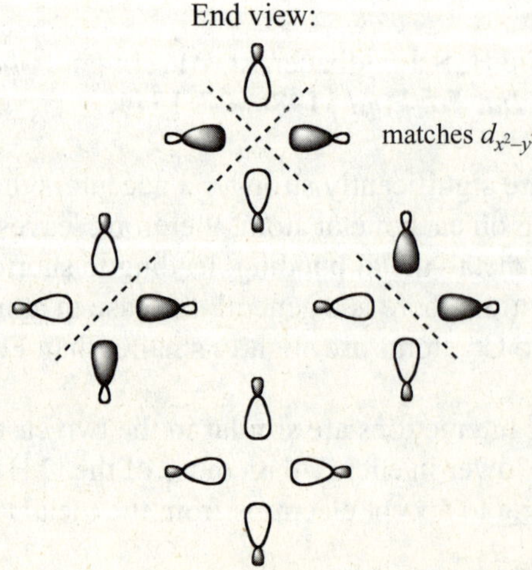

15.16 Cotton's explanation is that removal of an electron effectively changes the oxidation number of Tc, causing the *d* orbitals in $[TcCl_8]^{2-}$ to be smaller than those of $[TcCl_8]^{3-}$. This reduces the orbital overlap, weakening the bonding in spite of the higher formal bond order. The change is small—about 3 pm. (See F. A. Cotton and G. Wilkinson, *Advanced Inorganic Chemistry*, 5th Ed., Wiley, 1988. p. 1090.)

15.17 As the reference describes, these ions are isostructural and are based on octahedra fused at one mutual face (occupied by three bridging chlorine atoms). The shortening of the Re–Re distance upon reduction is attributed chiefly to two factors. As the complex is reduced, the net positive charge on each metal is reduced (the oxidation state changes from 4 to 3.5), which reduces the metal–metal repulsion. In addition, with the reduction of the metals, the metal *d* orbitals expand, enabling more effective overlap and stronger bonding.

15.18 This example is similar to the one presented in Figure 15.10. In compound **1** there are six DTolF ligands, each with a charge of 1–. Adding the charges of the bridging hydroxides, the total charge of the ligands in this complex is 8–. Consequently, the total charge on the Mo atoms must be 8+, or a charge of 4+ per Mo_2 unit. Each Mo^{2+} has 4 *d* electrons, giving 8 *d* electrons per Mo_2. Mo_2^{4+}, with its 8 *d* electrons, has an Mo–Mo bond order of 4 (Figure 15.10).

In compound **2** there are again six DTolF ligands. The bridging O^{2-} ligands result in a total ligand charge of 10–; consequently, the four molybdenums must carry an average charge of 2.5+, or 5+ per Mo_2 unit. There are now just 7 *d* electrons per Mo_2. From Figure 15.10 we can see that Mo_2^{5+}, with 7 *d* electrons, corresponds to a bond order of 3.5. The lower Mo–Mo bond order in **2** results in a longer bond.

15.19 **a.**

O_h	E	$8C_3$	$6C_2$	$6C_4$	$3C_2$	i	$6S_4$	$6S_6$	$3\sigma_h$	$6\sigma_d$
$\Gamma\ (s, p_z)$	6	0	0	2	2	0	0	0	4	2
A_{1g}	1	1	1	1	1	1	1	1	1	1
T_{1u}	3	0	–1	1	–1	–3	–1	0	1	1
E_g	2	–1	0	0	2	2	0	–1	2	0

Γ fits either the *s* or the p_z orbitals.

b. It can be seen from the table above that $\Gamma = A_{1g} + T_{1u} + E_g$. This can also be worked out by the more elaborate methods used in Chapter 4.

c. The reducible representation for the p_x and p_y orbitals and the components of this representation are shown in the table below. Figure 15.13 shows the three T_{1u} and the three T_{2g} group orbitals required (which also include some p_z contribution from two of the boron atoms). The T_{1g} and T_{2u} representations are for antibonding orbitals.

O_h	E	$8C_3$	$6C_2$	$6C_4$	$3C_2$	i	$6S_4$	$6S_6$	$3\sigma_h$	$6\sigma_d$	
$\Gamma_{x,y}$	12	0	0	0	–4	0	0	0	0	0	
T_{1g}	3	0	–1	1	–1	3	1	0	–1	–1	(R_x, R_y, R_z)
T_{1u}	3	0	–1	1	–1	–3	–1	0	1	1	(x, y, z)
T_{2g}	3	0	1	–1	–1	3	–1	0	–1	1	
T_{2u}	3	0	1	–1	–1	–3	1	0	1	–1	

15.20 The number of orbitals of each type can be obtained by analogy with the results for $B_6H_6^{2-}$ on page 608:

2 valence atomic orbitals of B combine to form:
- 15 bonding orbitals ($2n + 1$) consisting of:
  - 8 framework MOs ($n + 1$)
    - 1 bonding orbital from overlap of *sp* orbitals
    - 7 bonding orbitals from overlap of *p* orbitals of B with *sp* hybrid orbitals or *p* orbitals of other B atoms
  - 7 B–H bonding orbitals (n)
- 13 nonbonding or antibonding orbitals

15.21
- **a.** $C_2B_3H_7 \rightarrow B_5H_9 \rightarrow B_5H_5^{4-}$ *nido*
- **b.** $B_6H_{12} \rightarrow B_6H_6^{6-}$ *arachno*
- **c.** $B_{11}H_{11}^{2-}$ *closo*
- **d.** $C_3B_5H_7 \rightarrow B_8H_{10} \rightarrow B_8H_8^{2-}$ *closo*
- **e.** $CB_{10}H_{13}^{-} \rightarrow B_{11}H_{14}^{-} \rightarrow B_{11}H_{11}^{4-}$ *nido*
- **f.** $B_{10}H_{14}^{2-} \rightarrow B_{10}H_{10}^{6-}$ *arachno*

15.22
- **a.** $SB_{10}H_{10}^{2-} \rightarrow B_{11}H_{13}^{2-} \rightarrow B_{11}H_{11}^{4-}$ *nido*
- **b.** $NCB_{10}H_{11} \rightarrow B_{12}H_{14} \rightarrow B_{12}H_{12}^{2-}$ *closo*
- **c.** $SiC_2B_4H_{10} \rightarrow B_7H_{13} \rightarrow B_7H_7^{6-}$ *arachno*
- **d.** $As_2C_2B_7H_9 \rightarrow B_{11}H_{15} \rightarrow B_{11}H_{11}^{4-}$ *nido*
- **e.** $PCB_9H_{11}^{-} \rightarrow B_{11}H_{14}^{-} \rightarrow B_{11}H_{11}^{4-}$ *nido*

15.23
- **a.** $B_3H_8(Mn(CO)_3) \rightarrow B_4H_8 \rightarrow B_4H_4^{4-}$ *nido*
- **b.** $B_4H_6(CoCp)_2 \rightarrow B_6H_8 \rightarrow B_6H_6^{2-}$ *closo*
- **c.** $C_2B_7H_{11}CoCp \rightarrow B_9H_{13}CoCp \rightarrow B_{10}H_{14} \rightarrow B_{10}H_{10}^{4-}$ *nido*
- **d.** $B_5H_{10}FeCp \rightarrow B_6H_{10} \rightarrow B_6H_6^{4-}$ *nido*
- **e.** $C_2B_9H_{11}Ru(CO)_3 \rightarrow B_{11}H_{13}Ru(CO)_3 \rightarrow B_{12}H_{14} \rightarrow B_{12}H_{12}^{2-}$ *closo*

15.24 **a.** Ge_9^{4-} has 40 valence electrons = $4n + 4$. Its classification is *nido*.

b. $InBi_3^{2-}$ has 3 + 3(5) + 2 = 20 valence electrons = $4n + 4$. Its classification is also *nido*.

c. Bi_8^{2+} has 8(5) – 2 = 38 valence electrons = $4n + 6$. It is an *arachno* cluster.

15.25 **a.**
$m = 1$ (a single polyhedron)
$n = 5$ (each B atom counts)
$o = 0$ (no bridging atoms)
$p = 2$ (2 missing vertices)
8 skeletal electron pairs

b.
$m = 2$ (2 polyhedra)
$n = 10$ (each B, C, and Co atom counts)
$o = 1$ (1 bridging atom, the cobalt)
$p = 2$ (2 missing vertices: the top part of the molecule is considered as a pentagonal bipyramid with the top vertex missing, and the bottom as an octahedron with the bottom vertex missing)
15 skeletal electron pairs

c.
$m = 2$ (2 polyhedra)
$n = 17$ (each B, C, and Fe atom counts)
$o = 1$ (1 bridging atom, Fe)
$p = 1$ (1 missing vertex, the top atom of the incomplete pentagonal bipyramid)
21 skeletal electron pairs

15.26 **a.** C_{2v}

b. C_{5v}, D_{5h}

c. $[Re_2Cl_8]^{2-}$: D_{4h}; $[Os_2Cl_8]^{2-}$: D_{4d}

d. D_{2h}

e. C_{5v}

f. D_{2h}, C_s

g. Te_6^{2+}: D_{3h} Ge_9^{4-}: C_{4v}

15.27 **a.** T_d

b. I_h

15.28 a. The group orbitals are derived primarily from the 3*p* orbitals of phosphorus, which collectively resemble the five π orbitals for C_5H_5 in Figure 13.22. Diagrams of these orbitals can be found in the second reference. (See also Z-Z. Liu, W-Q. Tian, J-K. Feng, G. Zhang, and W-Q. Li, *J. Phys. Chem. A*, **2005**, *109*, 5645.)

Atomic orbitals on transition metals suitable for interaction (assuming metal centered below P_5 plane)

lowest energy group orbital: s, p_z, d_{z^2}
1-node degenerate pair: p_x, d_{xz}, p_y, d_{yz}
2-node degenerate pair: d_{xy}, $d_{x^2-y^2}$

b. The molecular orbitals of P_5^- are lower in energy than the similar orbitals of $C_5H_5^-$, giving rise to a generally stronger ability of P_5^- to π accept. For example, the energy match between 2-node orbitals and metal *d* orbitals may be closer in P_5^- complexes than in the case of ferrocene (Figure 13.28), enabling stronger interaction. An example of such an interaction would be between an empty 2-node orbital of P_5^- and a $d_{x^2-y^2}$ orbital of a metal:

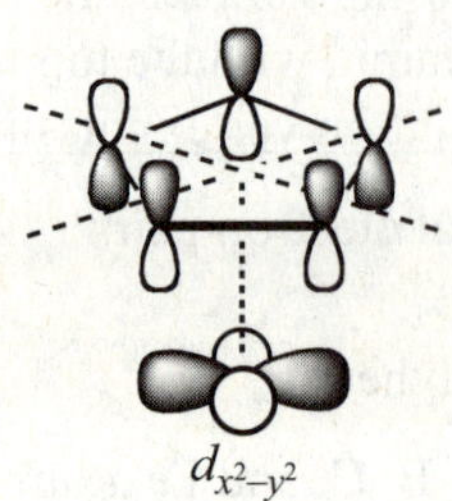

c. The reference provides an energy level diagram of the molecular orbitals of $[(\eta^5\text{-}C_5H_5)_2Ti]^{2-}$. In analyzing the orbitals it is important to see how the π orbitals of P_5^- ligands match up with *d* orbitals on Ti and how the resulting shapes illustrate how the lobes of these interacting orbitals merge (in the bonding orbitals) and how nodes are formed between them (in the antibonding orbitals).

15.29 The three types of interaction, σ, π, and δ, should be evident in the orbitals generated. In addition to the bonding interactions (see Figure 15.8) there should be matching antibonding interactions, with a nodal plane bisecting the metal–metal bond. The relative energies of the molecular orbitals should be similar to those on the right side of Figure 15.9, depending on the level of sophistication of the calculations used.

15.30 In addition to the σ, π, and δ interactions between *d* orbitals, shown in Figure 15.8, a σ interaction could occur between *s* orbitals of the two atoms:

s + *s* → σ

(In a chromium atom the valence 4*p* orbitals are empty, so interactions between them need not be considered.) The extent of the calculated interactions between the atomic orbitals is likely to differ significantly. Consequently, even though there may be six possible orbital interactions, the strength of such bonding is calculated to be significantly weaker than would be expected in a true "sextuple" bond. It is suggested that for this exercise the Cr–Cr distance be set at 166 pm, the equilibrium distance reported in the references.

15.31 **a.** The A_{1g} orbital should have a very large, nearly spherical lobe in the center of the cluster and six smaller, also nearly spherical lobes, on the outside of the borons (centered on the hydrogens).

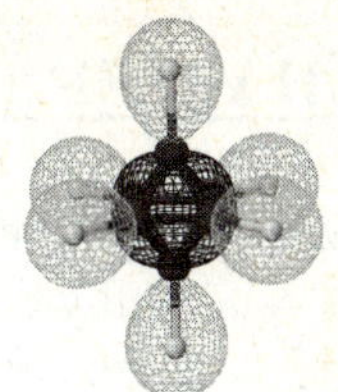

b. The T_{1u} orbitals should each have two regions where lobes of *p* orbitals on four B atoms merge, plus additional lobes, nearly spherical, centered on opposite hydrogens.

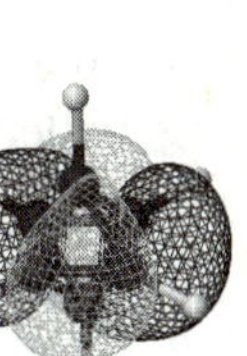

c. The T_{2g} orbitals should show how the lobes of adjacent *p* orbitals merge. The result should have four lobes, somewhat similar in appearance overall to a *d* orbital.

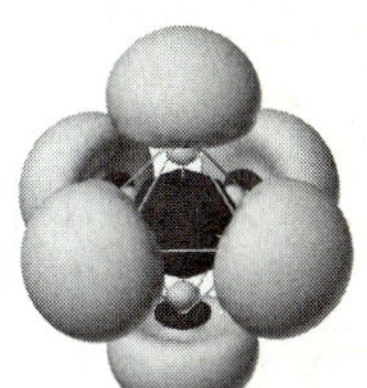

15.32 The orbitals are naturally much more complex, with 167 orbitals and 270 electrons. The T_{1u} and T_{2g} orbitals shown are from a Scigress Extended Hückel calculation.

a. The A_{1g} orbital should be similar to that for $B_6H_6^{2-}$ in problem 15.31, with one large lobe in the center and smaller *d* orbital lobes on each of the Ru atoms. It is the HOMO, antibonding in symmetry, and the only orbital near this energy that involves carbon orbitals.

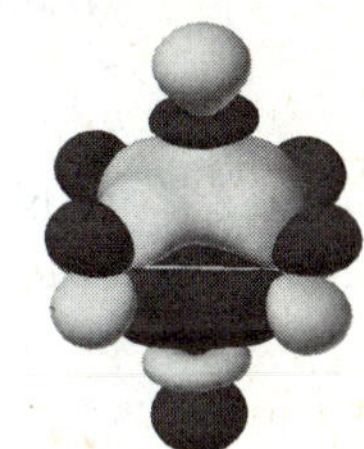

b. The T_{1u} orbitals should also be similar to the T_{1u} orbitals of $B_6H_6^{2-}$. There should be two large lobes each derived primarily from *d* orbitals on four Ru atoms and a *p* orbital of the central carbon, plus lobes on opposite Ru atoms (the other Ru atoms). There may also be fragment lobes of *d* orbitals on the Ru atoms.

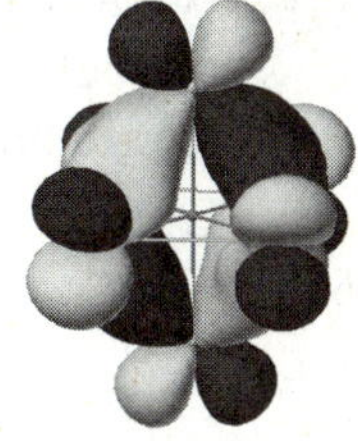

c. The T_{2g} orbitals are not directly involved in bonding with the carbon, but they do strengthen the cluster framework. The atomic orbitals should interact in sets of four, similarly to the T_{2g} orbitals in $B_6H_6^{2-}$, but some of the distinctive features of *d* orbitals should also be observable in the ruthenium cluster.

CHAPTER 16: BIOINORGANIC AND ENVIRONMENTAL CHEMISTRY

16.1 Methylation reactions can come from methylcobalamin, in which the methyl group is transferred to another substrate. Whether it is as CH_3^-, $\cdot CH_3$, or CH_3^+ is uncertain, but free radicals have been detected in reaction mixtures of this type.

$$LCoH_3 \longrightarrow LCo(I)^- + CH_3^+$$
$$\downarrow HSR$$
$$LCo(I)^- + H_3CSR + H^+$$

The source of the methyl groups is uncertain; perhaps from methane generated by bacterial action under anaerobic conditions.

Isomerization reactions are thought to require transfer of a hydrogen from the substrate to the 5′ carbon of the 5′-deoxyadenosyl group, followed by rearrangement of the substrate and transfer back of the hydrogen. Dehydration of 1,1-diols does not appear to require coenzyme B_{12}.

CH_2R–Co + CH_2R'–CH_2R'' ⟶ $\cdot CH_3R$ Co + $\cdot CHR'$–CH_2R''

⟶ $\cdot CH_3R$ Co + $CHR'R''$–$\cdot CH_2$ ⟶ CH_2R–Co + $CHR'R''$–CH_3

$R'H_2CCH_2R''$ ⟶ $R'R''HCCH_3$

CH_2R is the 5′-deoxyadenosyl group; there may be additional R groups on the substrate. Similar reactions can be written assuming ionic intermediates.

16.2 At high oxygen pressure, as in the capillaries of the lungs, hemoglobin picks up a full load of four O_2 molecules because of the cooperative nature of the bonding. At low oxygen pressure, as in the capillaries of the extremities, hemoglobin starts to release O_2, and the cooperative nature of the bonding results in release of all four. Myoglobin serves as a more passive storage device, being less sensitive to changes in O_2 concentration in the midrange. When oxygen is used in reactions, the concentration drops and more is released from the myoglobin. In a resting state, with low demand, it can load up from the higher concentration generated by the hemoglobin release of O_2.

16.3

[Resonance structures: Fe–O–N(H)–C(CH₃)=O⁺···Fe ⟷ Fe–O–N⁺(H)=C(CH₃)–O···Fe]

16.4 Hemoglobin contains Fe(II) in the high spin state. When paramagnetic O_2 binds, the result is a diamagnetic complex. Regardless of the description of the bonding, there must be pairing of the electrons of the iron and of O_2 in the resulting molecular orbitals. Many different suggestions have been made, with Fe(II)-O_2 seeming most natural from the starting materials. The structure proposed by Ochiai and shown in Figure 16.4 uses both π

antibonding orbitals of O_2 to form the molecular orbitals. Others have suggested that Fe(III)-O_2^- is more likely, in part because of the oxidizing power of the complex. In molecular orbital terms, all of these are essentially the same, since the result is a set of molecular orbitals, rather than specific components. Mössbauer spectra are influenced by the concentration and symmetry of electrons very close to the iron atom, and indicate the oxidation state of iron. Spectra measured by Tsai, Groves, and Wu show two different electronic environments. They interpret this as being an equilibrium 1:2 mixture of Fe^{2+}–O_2 and Fe^{3+}–O_2^-, with Fe^{2+} increasing at higher temperatures, and the energy levels shown below.

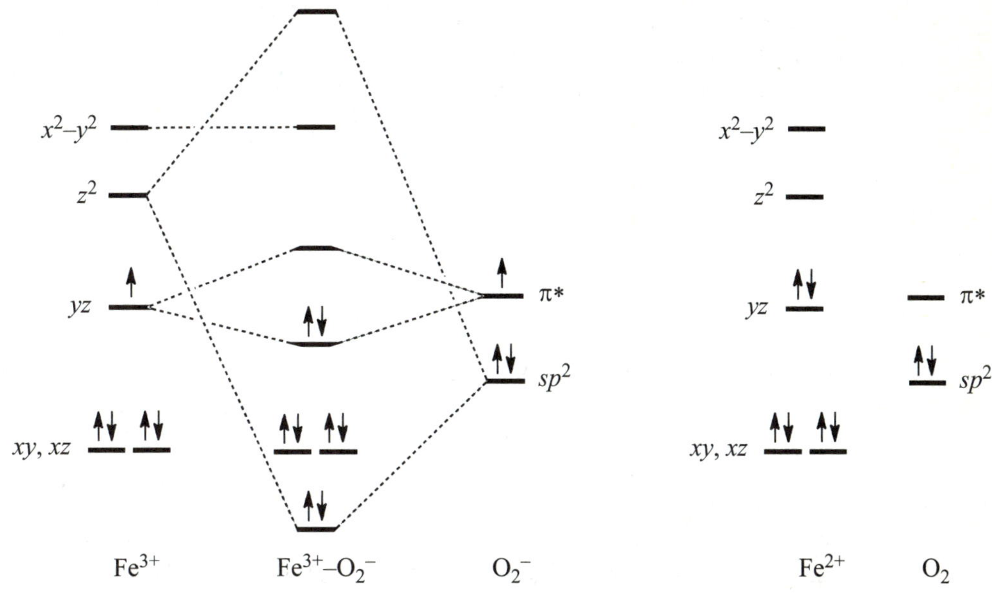

16.5 $\lambda = 395 \times 10^{-9}$m $\quad \bar{\nu} = 25{,}300\ cm^{-1} \quad E = 25{,}300\ cm^{-1} \times 11.96 \times 10^{-3}\ kJ\ mol^{-1}/cm^{-1}$
$= 303\ kJ\ mol^{-1}$

$\lambda = 242 \times 10^{-9}$m $\quad \bar{\nu} = 41{,}300\ cm^{-1} \quad E = 41{,}300\ cm^{-1} \times 11.96 \times 10^{-3}\ kJ\ mol^{-1}/cm^{-1}$
$= 494\ kJ\ mol^{-1}$

NO_2 has a bond order near 1.5, O_2 has a bond order of 2. The results here fit the bond order.

16.6 Methyl cobalamin is described as a Co(II) compound, forming Co(I) if CH_3^+ is formed, Co(II) if $\cdot CH_3$ is formed, and Co(III) if CH_3^- is formed. All three possibilities have been suggested. As Co(III), methylcobalamin would be a low-spin, nearly octahedral species. After dissociation of CH_3 with any of its three charges, the following electronic structures are possible.

Depending on the separation of the two higher levels, the Co(I) species could have two unpaired electrons, one in each (d_{z^2} and $d_{x^2-y^2}$).

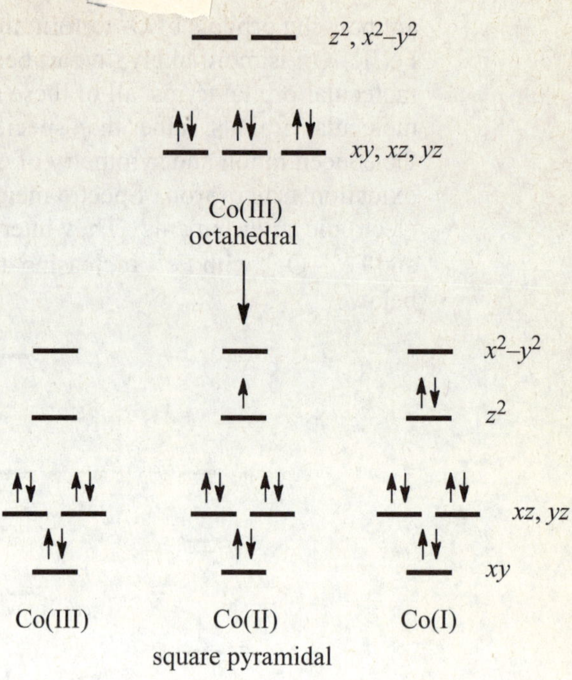

16.7 The EDTA-Pb^{2+} complex and similar complexes for other heavy metals are stable enough to work effectively in removing the heavy metals from the body. The danger is in removing useful metals as well (Ca^{2+}, Mg^{2+}, other transition metals). Stability constants for many of these metals are also high, so all may be removed. Fortunately, the bulk of these other metals is tied up in other compounds and is relatively slow to be released. Since they must be released from other tissues into the blood, their removal is likely to be slow compared with toxic metals ingested only a short time before. If the treatment has been delayed, the problem can be more serious because the toxic metal ions are more likely to be also tied up in tissues, and their rate of removal may be similar to that of the essential metal ions.

16.8 If the synthetic blood does not have the ability to absorb excess NO, its use can lead to dilation of blood vessels and low blood pressure, leading to inflammatory disease and degeneration of nerve tissue. On the other hand, if synthesis of NO is decreased because the appropriate enzymes are not present in the blood vessels, some of the same transport compounds are needed to maintain a level adequate to prevent the effects of low NO: hypertension, arteriosclerosis, impotence, and susceptibility to infection.

16.9 **a.** This structure, shown below, is the pyranopterindithiolate ligand; it has also been called molybdopterin and pterindithiolenc.

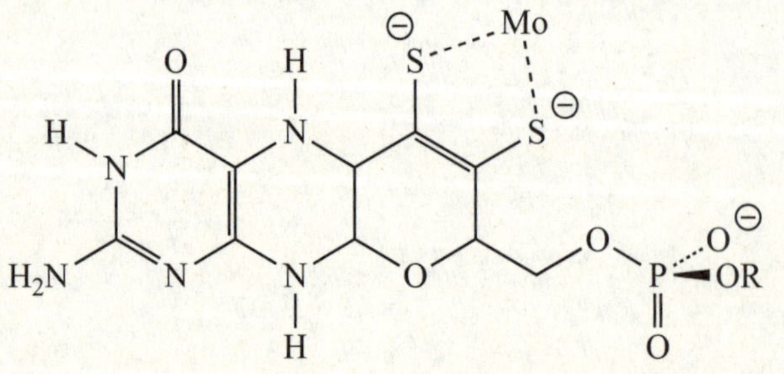

O H Mo S S H N N H₂N N N O O O P OR O H

b. Representative active sites in these families are shown below (Cys = cysteine, Ser = serine). Examples of related structures are shown in the references.

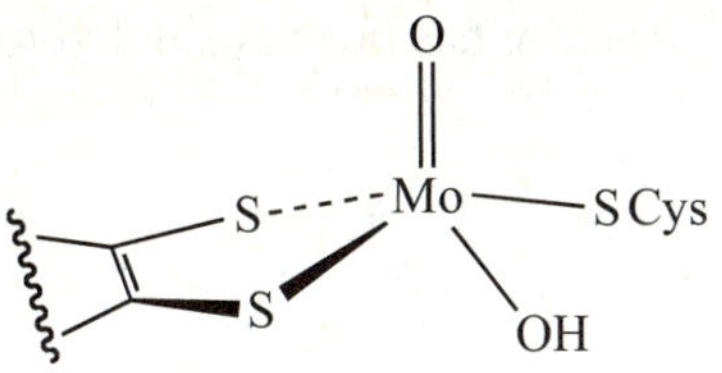

Sulfite oxidase

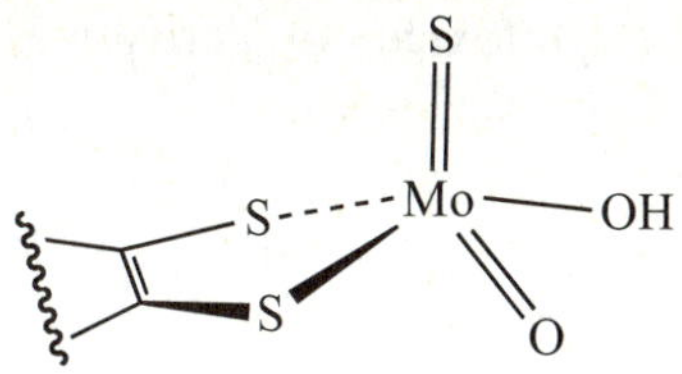

Xanthine oxidase

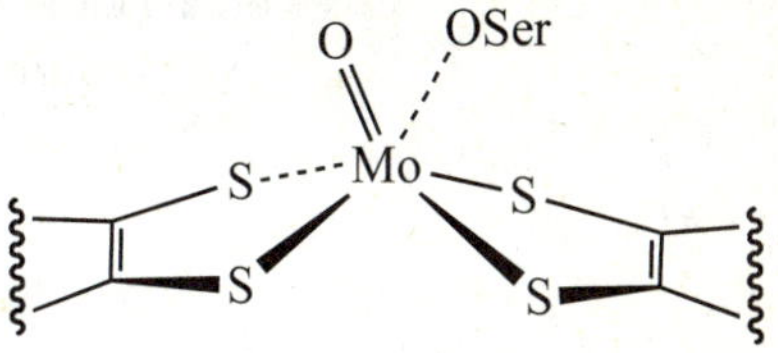

DMSO reductase (oxidized form)

c. The range of reactions catalyzed is broad; specific examples are cited early in the 2004 reference. (See also M. J. Romão, *Dalton Trans.*, **2009**, 4053 for additional information on structures and reactivities of molybdenum and tungsten enzymes.)

16.10 a. The formate dehydrogenase family is structurally similar to the DMSO reductase family of molybdenum enzymes. (See also the references cited in problem 9c.)

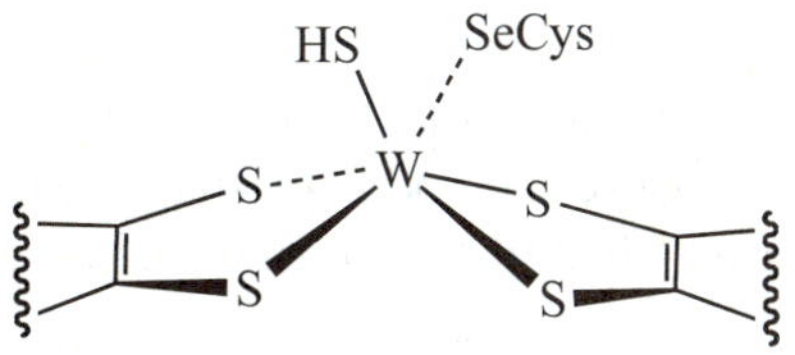

Formate dehydrogenase

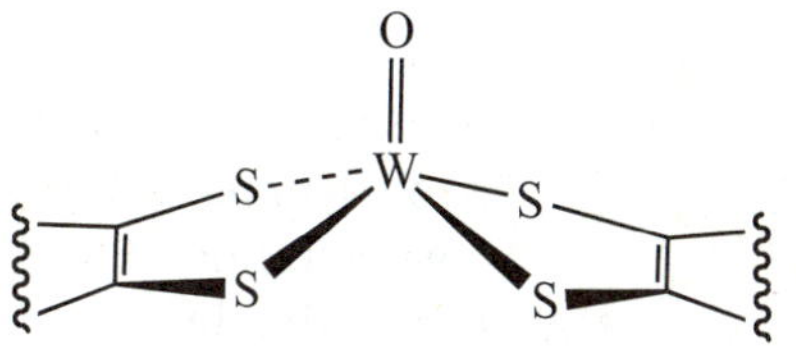

Aldehyde oxidoreductase

b. Aldehyde oxidoreductase enzymes catalyze the oxidation of aldehydes to carboxylic acids, and formate dehydrogenase enzymes catalyze the oxidation of formate to carbon dioxide.

c. Tungsten-containing enzymes occur most frequently in archaea (archaebacteria), which have been viewed as most similar to the types of life that existed in early geologic times, and these enzymes are not known at all in eukarya, in which molybdenum-containing enzymes play essential roles. The 2009 reference provides a brief discussion of how the occurrence of tungsten enzymes is much more complex than the notion of relative age.

16.11 Cisplatin and epidermal growth factor were attached to single–walled carbon nanotubes, which entered head and neck squamous cells in mice. Green and red emitting quantum dots were also attached and were used to monitor uptake of the derivatized nanotubes by the cells.

16.12 Ferrocifen derivatives have shown activity against human breast cell lines. These contain a ferrocenyl group that has been attached to a structure similar to that of tamoxifen, a drug that has been used widely for the treatment of patients who have hormone dependent breast cancer. The cytotoxicity of derivatives of ferrocifen has been attributed to the oxidation of Fe^{2+} to Fe^{3+}.

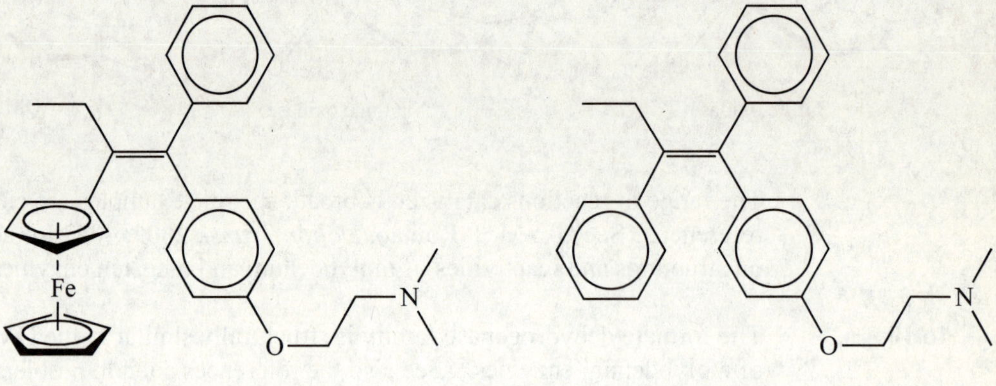

Fe O N

Ferrocifen

O N

Tamoxifen

16.13 **a.** These maxima and minima result from the annual cycle of plant activity. When photosynthesis is most active, plants use up more CO_2 than is formed, and the concentration of this gas decreases; when photosynthesis is less active, the production of CO_2 is dominant.

b. Graphing data over long time periods shows an increasing rate of CO_2 growth, with the current rate of increase more than double the rate observed during 1959–1969, the first decade of observations recorded at Mauna Loa. There tends to be considerable fluctuation from one year to the next, so considering a long time period is recommended. Also, comparisons with other sampling sites can provide perspective on the degree to which an increase in CO_2 concentration is a worldwide phenomenon.

c. Predictions can differ considerably, depending on the underlying assumptions and the mathematical techniques applied. It can be instructive to use early data (for example, from 1959–1984) to see how accurately the current CO_2 concentration might be predicted based on different models.

16.14 Useful references for this exercise can also be found using electronic search tools such as Web of Science and SciFinder.